The Light
Eaters

The Light Eaters

Zoë Schlanger

4th ESTATE • London

4th Estate
An imprint of HarperCollins*Publishers*
1 London Bridge Street
London SE1 9GF

www.4thestate.co.uk

HarperCollins*Publishers*
Macken House
39/40 Mayor Street Upper
Dublin 1
D01 C9W8, Ireland

First published in Great Britain in 2024 by 4th Estate

First published in the United States by Harper, an imprint of
HarperCollins*Publishers*, in 2024

1

Typeset in Ashbury

Printed and bound in the UK using 100% renewable electricity at CPI Group (UK) Ltd

This book contains FSC™ certified paper and other controlled sources
to ensure responsible forest management.

For more information visit: www.harpercollins.co.uk/green

To Anne and Jeff, who see the big meaning in small things.

They can eat light, isn't that enough?

—*Timothy Plowman, ethnobotanist*

Contents

The Light Eaters

Prologue

I am walking along a dim path. Thick hillocks of moss undulate fuzzily around me. I look up, and am dwarfed by pillars of dank and slimy wood. The earth below me is damp, has give. A sign on the path tells me to be alert for aggressive elk in the area. I see no elk, I keep walking. Plumes emerge, sword ferns with their curled fiddleheads the size of a baby's fist covered in velvety auburn hair, the unexpected prequel to the arching fronds that will fountain out above them like peacock feathers. Moss drips in long fingers from branches overhead. Fungi arc skyward from a downed tree. Everything seems to strain upward and downward and outward at once.

I intrude on all this, but no one notices. All things here are so thoroughly absorbed into their own living that I am like an ant slipping discreetly through a sponge. The lichens crawling up the base of trees curl the edges of their disklike bodies up, catching drops as they receive a new day and another chance to grow.

I am in the Hoh Rain Forest in the Pacific Northwest, and everywhere is a sense of secrets. And for good reason. For everything that science does know about what, biologically, is going on here, there is so much more it cannot yet explain. All around me are complex adaptive systems. Each creature is folded into layers of interrelationship with surrounding creatures that cascade from the largest to the smallest scale. The plants with the soil, the soil with its microbes, the microbes with the plants, the plants with the fungi, the fungi with the soil. The plants with the animals that graze on them and pollinate them. The plants with each other. The whole beautiful mess defies categorization.

Thinking about this, I am reminded of the concepts of yang and yin, the philosophy of opposing forces. We know that the forces that shape life are in constant flux. The moth that pollinates the flower of a plant is the same species that devours the plants' leaves when it is still a caterpillar. It is not, then, in the plant's interest to completely destroy the grazing caterpillars that will metamorphose into the very creatures it relies on to spread its pollen. But likewise the plant cannot bear total leaf destruction; without leaves, the plant can't eat light, and it will die. So after a while the besieged plant, having already lost some limbs and therefore showing tremendous restraint, judiciously begins to fill its leaves with unappetizing chemicals. At least most of the caterpillars will have eaten enough to survive, metamorphose, pollinate. Everyone in this situation comes within a hair of death to ultimately flourish. This is the push and pull of interdependence and competition. At the grand scale, no one seems to have yet won. All parties are still here, animals, plants, fungi, bacteria. What we end up with is a sort of balance in constant motion. All of this pushing and pulling and coalescing, as I have come to understand, is a sign of tremendous biological creativity.

How to get our minds around all of that complexity is the shared professional problem of science and philosophy, but also of every person who's stopped to wonder. All that roiling life that won't stand still long enough for us to get a good look. Narrowing our focus to only plants seems at first sensical; that should be easier, being one thing. But that quickly proves naive. Complexity lives at every scale.

Journalists in my line of work tend to be focused on death. Or the harbingers of it: disease, disaster, decline. That is how climate journalists mark time as the earth passes benchmark after grim benchmark on its way into the foreseen crisis. There is only so much of this that one person can take. Or perhaps my tolerance was thin and easily worn out after years of focus on droughts and floods. In recent years I'd begun to feel numb and empty. I needed some of the opposite. What, I wondered, is the opposite of death? Creation, perhaps. A sense of becomings instead of endings. Plants are that, given as they are to continuous growth. They'd soothed me all my life, long before studies came out confirming what we already knew: that time spent among plants can ease the mind better than a long sleep. Living in a dense city, I'd walked in the park under a canopy of yews and elms when I needed to clear my head; I'd spent long minutes gazing at the new leaves forming on my potted philodendrons when my nerves were fried. Plants are the very definition of creative becoming: they are in constant motion, albeit slow motion, probing the air and soil in a relentless quest for a livable future.

In the city, they seemed to make a home in the least suitable places. They burst from cracks in crumbling pavement. They climbed the chain link fence at the edges of garbage-strewn lots. I privately delighted as I watched a tree of heaven—loathed as an invasive species in the Northeast—emerge out of a split on my stoop and grow nearly to the size of a two-story building in a single season. Privately, because I knew this was seen as a hellish species in New York, in part because it injects poisons into the ground around its roots to prevent anything else from growing nearby, thereby securing its patch of sun; delighted, because this seems devilishly brilliant. When my neighbor cut the tree down with a machete late in the season, I understood. Still, I regarded its stump with admiration on my way out of the house every morning. It was already sprouting new green protrusions. You have to respect a good hustle.

So plants seemed like the right place to land my weary apocalyptic attention. Surely they would refresh me. But I soon learned they would do more than that. Plants have, over the course of years of obsession, transformed my understanding of what life means, and what its possibilities are. Now as I gaze around the Hoh Rain Forest I see more than a soothing wash of green. I see a masterclass in living to one's fullest, weirdest, most resourceful potential.

To start with, a life spent constantly growing yet rooted in a single spot comes with tremendous challenges. To meet them, plants have come up with some of the most creative methods for surviving of any living thing, us included. Many are so ingenious that they seem nearly impossible for an order of life we've mostly relegated to the margins of our own lives, the decoration that frames the theatrics of being an animal. Yet there they are all the same, these unbelievable abilities of plants, defying our anemic expectations. Their way of life is so astonishing, I will soon learn, that no one yet really knows the limits of what a plant can do. In fact, it seemed that no one quite knows what a plant really is.

This is of course a problem for the scientific field of botany. Or it's the most exciting thing to happen to it in a generation, depending on how comfortable you feel with seismic shifts in what you once thought to be true. Now I was hopelessly intrigued. Controversy in a scientific field tends to be a harbinger of something new, some new understanding of its subject. In this case, the subject was all of green life itself. I began to direct my growing interest entirely toward emerging thought in plant science. The more botanists uncovered the complexity of forms and behaviors of plants, the less the traditional assumptions about plant life seemed to apply. The scientific field was eating itself alive with contradictions, points of contention multiplying as fast as the mysteries. But something in me was attracted to this lack of neat answers, as I suspect many of us are. Who doesn't feel both drawn to and repulsed by the unknown?

This book will take up these new epiphanies in plant science, and the struggle along the way over how new scientific knowledge is made. Rarely does one get a glimpse of a field in true turmoil, debating the tenets of what it knows, about to birth a new conception of its subject. We will also consider a daring question currently being hotly debated in labs and academic journals: Are plants intelligent? Plants don't have brains, as far as anyone can tell. But, some posit, they should be considered intelligent regardless, based on the remarkable things they can do. We determine intelligence in ourselves and certain other species through inference—by observing how something behaves, not by looking for some physiological signal. If plants can do things we consider indications of intelligence in animals, this group says, then it's illogical, a sign of unreasonable zoocentric bias, not to use that language for them. Others go farther, suggesting that plants may be conscious. Consciousness is perhaps the least understood phenomenon in human beings, let alone other organisms. But a brain, this camp says, may be but one way to build a mind.

Other botanists are more circumspect, unwilling to apply what they see as distinctly animal-centric notions to plants. Plants, after all, are their own clade of life, with an evolutionary history that swerved away from our own long ago. Painting them with our concepts of intelligence and consciousness does a disservice to their essential plantness. We'll meet this camp of scientists too. Yet no one I met—not a single botanist—was anything less than agape with wonder at what they were learning that plants are capable of. Thanks to new technology, scientists in the last two decades have gained incredible new powers of observation. Their findings are reshaping the meaning of "plant" before our very eyes.

Regardless of what we think of plants, they continue to surge upward, toward the sun. In this ruined global moment, plants offer a window into a verdant way of thinking. For us to truly be part of this world, to be awake to its roiling aliveness, we need to understand plants. They suffuse our atmosphere with the oxygen we breathe, and they quite literally build our bodies out of sugars they spin from sunlight. They made the ingredients that first allowed our lives to blink into existence at all. Yet they are not merely utilitarian supply machines. They have complex, dynamic lives of their own—social lives, sex lives, and a whole suite of subtle sensory appreciations we mostly assume to be only the domain of animals. What's more, they sense things we can't even imagine, and occupy a world of information we can't see. Understanding plants will unlock a new horizon of understanding for humans: that we share our planet with and owe our lives to a form of life cunning in its own right, at once alien and familiar.

In the Hoh Rain Forest, a bigleaf maple stretches out above me. Its trunk is sheathed entirely in licorice ferns and lungwort and spike moss, giving the impression of a tree wearing a Grinch suit. Only a few ridges of the tree's bark are visible, rising through the green fuzz like a mountain range above a mat of thick woods, like the Olympic peaks that pierce the evergreen forests just east of here. I lean in, looking closer. The green suit is a world within a world, the little tufts and fronds replicating the structure of a forest at tiny scale. Three-leafed oxalis and feathery stairstep moss coat the ground. I am lost in their world, taken into it. Then again, we've all been lost in it a long time, unaware of its true machinations. This seems imprudent. I wanted to know, so I went out and looked.

Chapter 1

———

The
Question
of Plant
Consciousness

What is a plant? You likely have an idea. You might be picturing a fat sunflower with its hubcap face and thick felted stalk, or the bean vine looped around a trellis in your grandmother's yard. Maybe, like me, you're eyeing the golden pothos hanging in your kitchen window that you should probably water. A known entity; the green of every single day.

Of course you're right, in the same way humans have been able to point to an octopus for all of history and call it an octopus. But we didn't know, until recently, that octopuses could taste with their arms, use tools, remember human faces, see their world far more sensitively than we can ours; that they have neurons distributed throughout their bodies like a multitude of disparate miniature brains. So, then, what is an octopus? Something much more than we ever imagined.

The answer is only beginning to dawn on us, and has already revolutionized our understanding of nonhuman intelligence in one crucial aspect: the octopus's branch on the evolutionary tree diverged from ours very early in the history of animal life. Our last common ancestor was likely a flatworm that trailed the ocean floor more than five hundred million years ago.* Up until this point, we've located intelligence in animals much evolutionarily closer to ourselves, like dolphins, dogs, and primates, our much more recent cousins. But we now know that powerful cunning can evolve completely independently from our own. A similar tectonic shift is happening with plants, only—for now—more quietly, in the labs and field sites of one of the least flashy disciplines within the life sciences. But the weight of this new knowledge is threatening to burst the walls of the container in which we place plants in our minds. Ultimately it may change how we think about life altogether.

So what is a plant? I certainly thought I knew. And then I began to talk to botanists.

SEVERAL YEARS AGO, I was an environmental reporter with a problem. Most of my work focused on two things: the steady advance of climate change, and the health effects of polluted air and water. In other words, I was writing about humanity's unrelenting death plod. After five or six years on that beat, a crawling sense of dread threatened to eclipse me. In response, I began to act strangely. I'd explain the latest IPCC report—the ones telling us we had very few years to stave off catastrophe—to my colleagues with an eerie sort of glee, awaiting their paled faces. I'd often spend a morning ingesting news about record-breaking wildfires and hurricanes and pivot seamlessly to office gossip by lunch. The compartmentalization became so total that I could no longer muster any emotional response to environmental cataclysm. Melting ice sheets in Greenland started to look like just another good story.

It was around then that I began to search, without realizing it, for

---

* For comparison, the last common ancestor of humans and dolphins was a land-dwelling mammal who lived something like 50 million years ago. Our last common ancestor with chimpanzees lived just 6 million years ago.

something in the natural sciences that felt wonderful and alive. I liked plants; I loved watching my night-blooming jasmine clamber up my window frame and my fiddle-leaf fig burst open three new leaves in a sudden growth spurt after months of no visible change. My apartment was a refuge of satisfying plant drama, better than the drama unspooling inside my computer. So, I thought, why not turn my reporter brain toward them? I began searching botany journals on my lunch breaks, using the same online portals I used to find climate papers, a system that lets journalists see the latest research before it is available to the public, on the condition that they not publish stories about it until its stipulated release date. The journals were full of fundamental discoveries about plants; uncovering the evolutionary origins of bananas and understanding, at last, why some flowers are slippery (to deter nectar-thieving ants). I felt like I was spying on a version of science from an earlier time; were there really this many fundamentals left to be uncovered? Two weeks into my new fascination, I learned that a complete fern genome had been sequenced for the first time, and a paper on it would be coming out soon. I didn't yet know how remarkable that was—ferns, being extremely ancient, can have up to 720 pairs of chromosomes, versus humans' mere 23, which explained why the genomic revolution took so long to reach them. I was immediately taken by the image of the fern that accompanied the embargoed scientific paper. It was a photograph of a tiny, scalloped plant atop a researcher's thumbnail, an azolla. It was so green it looked lit from within. I was in love.

Azolla filiculoides, or just azolla for short, is one of the world's smallest ferns, and has grown in wet places for millennia. As is generally the rule with plants, it is unwise to mistake size for complexity. Roughly fifty million years ago, when the earth was a much warmer place, azolla began growing over the Arctic Ocean in vast fern blankets. For the next million years they absorbed so much CO_2 that paleobotanists believe they played a crucial role in cooling the planet, and some researchers are looking into whether they could help do that again.

The azolla performs another miraculous trick still; some one hundred million years ago, it evolved a specialized pocket in its body to house a packet of cyanobacterium that fixes nitrogen. The air around us is nearly 80 percent nitrogen, and every life form, including ours, needs it to manufacture nucleic acids, the building blocks of all life. But in its atmospheric form, it's entirely out of our reach. Nitrogen, nitrogen everywhere, and not a single molecule that we can use. In a humbling twist, plants rely wholly upon bacteria that know how to re-combine nitrogen into forms the plant—and all of us who get our ni-trogen from plants—can use. And so azolla morphed itself into a hotel for this bacterium. The tiny fern feeds the cyanobacteria the sugars it needs, and the cyanobacteria busy themselves with transforming ni-trogen. Farmers in China and Vietnam took note of this and have been grinding azolla into their rice paddies for centuries.

I sought out fern guidebooks and fern lore. I was charmed by my own rapaciousness, which had been activated like this only a few times in my life. I was so enamored that I got the tiny azolla tattooed on my left arm. Journalists are notorious generalists, interested briefly and drenchingly in many things they soon leave behind. But this was, I thought, what it must be like to be *taken* with something. I had questions, suddenly, about this commonest group of plants that had sprouted seemingly without fanfare. They changed the world. What else did I not know?

As part of this inquiry I bought and devoured *Oaxaca Journal*, a slender volume of Oliver Sacks's observations from a fern expedition to southwestern Mexico he took with a bus full of dedicated amateur pteridologists, all of them from the New York chapter of the American Fern Society. The expedition was co-led by Robbin C. Moran, a forty-four-year-old curator of ferns at the New York Botanical Garden, who brought them all around the state of Oaxaca. At one point, after days of visiting villages and landscapes, marveling at produce in the markets and dye vats of red cochineal, and of course all manner of liverworts and ferns, Sacks has a moment that can only be described as rapture. The afternoon sun is falling strong and slant against high stalks of corn. An older gentleman, a botanist and specialist on Oaxacan agri-culture, is standing beside the corn. Sacks acknowledges the numinous moment—the briefest flicker—with barely half a sentence, but it stung me immediately as true.

> ... the tall corn, the strong sun, the old man, become one. This is one of those moments, indescribable, when there is a sense of intense reality, an almost preternatural reality—and then we are descending the trail to the gate, reboarding the bus, all in a sort of trance or daze, as if we had had a sudden vision of the sacred, but were now back in the secular, everyday world.

The experience of flashes of the eternal, the real, the gestalt, runs like a thread throughout naturalist literature. I wasn't the only one who had been taken like this before. In *Pilgrim on Tinker Creek*, the writer Annie Dillard has a similar moment in front of a tree, watching light pour through its branches. A flash of the real. Almost as soon as she realizes she is having it, it is gone, but it leaves her with the awareness of a sort of open-plan attentiveness that can be accessed in snippets, and which might be a more direct observation of the world than the usual everyday version.

As I read more books about plants and their enraptured naturalists after work and into the early morning, I began to find these moments sprinkled everywhere. In *The Invention of Nature*, Andrea Wulf's biography of the famous nineteenth-century naturalist Alexander von Humboldt, I learned he'd had it too. Von Humboldt wondered aloud why being in the outdoors evoked something existential and true. "Nature everywhere speaks to man in a voice that is familiar to his soul," he wrote; "Everything is interaction and reciprocal," and therefore nature "gives the impression of the whole." Humboldt went on to introduce the European intellectual world to the concept of the planet as a living whole, with climatic systems and interlocking biological and geological patterns bound up as a "net-like, intricate fabric." This was Western science's earliest glimmer of ecological thinking, where the natural world became a series of biotic communities, each acting upon the others.

Something about reading botany papers gave me snatches of that feeling, glimpses of a sort of whole I couldn't yet fully articulate. I had the sense I was uncovering large gulfs in my knowledge. How long had I spent in the presence of plants while knowing next to nothing about them? I felt a curtain drawing back little by little onto a parallel universe. I knew it was there now, but not yet what it held.

I signed up for a fern-science class at the New York Botanical Garden, taught by none other than the Moran from Sacks's expedition, no longer forty-four but youthful just the same. (I would come to learn that the botany world comprises a cast of recurring characters with storylines between them, some amiable, some not.) We learned how to identify ferns, about their basic structure, and about the more idiosyncratic species; a resurrection fern grows on the branches of oak trees, and can almost completely dehydrate in times of drought, shrinking down into a dead-looking crisp. It can remain in its dried state for more than a hundred years and still fully rehydrate. Tree ferns can grow more than sixty-five feet high, and other ferns are minuscule fertilizer factories, like the tiny azolla. And then there's bracken fern, which makes cows who dare to graze it bleed internally to death. "Just a totally cruel fern," Moran said.

I learned that ferns are much, much older evolutionarily than flowering plants. They appeared on the scene even before evolution dreamed the concept of seeds; they reproduce without them. Days later, during lunch hours spent reading about ferns, falling desperately into true obsession, I learned that this lack of seeds boggled Europeans for centuries. All plants had seeds; it was key to their sexual reproduction, or so medieval people thought. If they couldn't find the fern's seeds, contemporary logic followed that they simply must be invisible. And because another dominant theory at the time suggested that the physical characteristics of plants were clues about what they could be used for, people believed finding these invisible seeds would grant humans the power of invisibility.

Actual fern sex turned out to be much weirder. First of all, they reproduce using spores, not seeds. But here's the kicker: they have swimming sperm. Before they grow into the leafy fronds we all know, they have a completely separate life as a gametophyte fern, a tiny lobed plant just one cell thick—not remotely recognizable as the fern it will later become. You'd miss them on the forest floor. The male gametophyte fern releases sperm that swim in water collected on the ground after a rain, looking for female gametophyte fern eggs to fertilize. Fern sperm are shaped like tiny corkscrews and are endurance athletes— they can swim for up to sixty minutes. You can watch them squiggle under a microscope.

The sperm alone isn't the most amazing thing about fern reproduction. In 2018, at the beginning of my infatuation, research was emerging to suggest that ferns compete with other ferns for resources by emitting a hormone that causes the sperm of neighboring fern species to slow down. Slower sperm means less of that species survives, so the sabotaging fern can enjoy more of whatever is scarce, be it water, sunlight, or soil.

Scientists were just beginning to wrap their heads around this fact. "This is brand-new," Eric Schuettpelz, a research botanist at the Smithsonian's National Museum of Natural History in Washington, DC, told me over the phone. Sabotaging sperm was evidently the cutting edge of fern science. "We know it's the plant hormone but no idea how it works," he said. How did a fern know it was beside some fern competition? How did it time its malevolent release? A fern researcher at Colgate University had just presented an early paper on the phenomenon at a botany conference that same month.

I let that sink in a minute: ferns can remotely interfere with other ferns' sperm. This was incredibly salty plant activity. I began to see where Moran was coming from. This also seemed remarkably brilliant. What else could plants do?

With that question in mind I began to focus my newfound lens on a relatively young area within plant science: plant behavior. The announcements of emerging research, I found, were peppered with plant behavior papers. This represented a new mental gate for me to walk through; that plants could be thought to behave at all was still an enchanting possibility. But several papers I found pushed the farthest edge of that concept farther: they suggested that plants might have a form of intelligence. I was intrigued, and skeptical. I wasn't the only one. As it turned out, the suggestion of plant intelligence had recently started an all-out war.

I'd come across this corner of the scientific world at a remarkably exciting time. In the last decade and a half a revival of plant behavior research had brought countless new realizations to botany, more than forty years after an irresponsible best-selling book nearly snuffed out the field for good. *The Secret Life of Plants*, published in 1973, captured the public imagination on a global scale. Written by Peter Tompkins and Christopher Bird, the book was a mix of real science, flimsy experiments, and unscientific projection. In one chapter, Tompkins and Bird suggested that plants could feel and hear—and that they preferred Beethoven to rock and roll. In another, a former CIA agent named Cleve Backster hooked up a polygraph test to his houseplant and imagined the plant being set on fire. The polygraph needle went wild, which would mean the plant was experiencing a surge in electrical activity. In humans, a reading like that was believed to denote a surge of stress. The plant, according to Backster, was responding to his malevolent thoughts. The implication was that there existed not only a sort of plant consciousness but also plant mind-reading.

The book was an immediate and meteoric success on the popular market, surprising for a book about plant science. Paramount put out a feature film about it. Stevie Wonder wrote the soundtrack. The first pressings of the album version were sent out scented with floral perfume. To its many astonished readers, the book offered a new way to view the plants all around them, which up until then had seemed ornamental, passive, more akin to the world of rocks than animals. It also aligned with the advent of New Age culture, which was ready to inhale stories about how plants were as alive as we are. People began talking to their houseplants, and leaving classical music playing for their ficus when they went out.

But it was a beautiful collection of myths. Many scientists tried to reproduce the most tantalizing "research" the book presented, to no avail. Cellular and molecular physiologist Clifford Slayman and plant physiologist Arthur Galston, writing in *American Scientist* in 1979, called it a "corpus of fallacious or unprovable claims." It didn't help that former CIA agent Backster, as well as Marcel Vogel, an IBM researcher who claimed to be able to reproduce the "Backster effect," believed that one had to develop an emotional rapport with a plant before any effect was possible. In their view, that explained away any inability of another lab to replicate the results. "Empathy between plant and human is the *key*," Vogel said, and "spiritual development is indispensable."

According to botanists working at the time, the damage that *Secret Life* caused to the field cannot be overstated. The twin gatekeepers of science funding boards and peer review boards—always conservative institutions—closed the doors. Over the following years, according to several researchers I spoke with, the National Science Foundation became more reluctant to give grants to anyone studying plants' responses to their environment. Proposals with so much as a whiff of inquiry into plant behavior were turned down. The money, whatever little there had been, dried up. Scientists who had pioneered the field changed course or left the sciences altogether.

But a select few held on, biding their time with other lines of inquiry, waiting for the tide to turn.

In the last decade and a half, it finally has. Funding for some plant behavior research began flowing again, though the grants were still hard to come by at first. Botany journals, though many were still edited by opponents of the plant intelligence field, began letting a trickle of these papers through. The change was likely the result of new technology like genetic sequencing and more advanced microscopes, which made it possible to come to previously outlandish conclusions with real rigor. Or perhaps the political mockery that followed the *Secret Life* debacle was just far enough in the rearview mirror. Many of the authors didn't use words like *intelligence* to describe what they found, but the results nonetheless suggested that plants were much more sophisticated than anyone had dared think.

Recently, as I came across in my reading, researchers had found promising indicators of memory in plants. Others found that a wide variety of plants are able to distinguish themselves from others, and can tell whether or not those others are genetic kin. When such plants find themselves beside their siblings, they rearrange their leaves within two days to avoid shading them. Pea shoot roots appeared to be able to hear water flowing through sealed pipes and grow toward them, and several plants, including lima beans and tobacco, can react to an attack of munching insects by summoning those insects' specific predators to come pick them off. (Other plants—including a particular tomato—secrete a chemical that cause hungry caterpillars to turn away from devouring their leaves to eat each other instead.) Papers probing other remarkable behaviors were growing from a trickle to a fairly robust stream. It seemed like botany was on the verge of something new. I wanted to stick around and watch.

BACK AT MY desk in the air-conditioned newsroom, I savored these small tears in the fabric of my day. Something about this renaissance in the study of plant behavior spoke to an earlier me. I was an only child for the first nine years of my life, until my brother was born. Even then a newborn wasn't much use to a nine-year-old girl, especially not one who truly believed she was an adult trapped in a child's body. Which is to say I was alone and predisposed to fantasy. Girls of that nature tend to build complex internal worlds that they proceed to drape like a blanket over the world around them. Adults who don't understand this disposition tend to call it melodramatic. But I resented that word, which implied that my version of reality couldn't be trusted. I was sure I was simply seeing things around me for what they actually were. In most cases, those things were trees and squirrels and sometimes rocks, and they were very much alive, alert to the world. Children are known to be inborn animists.

Noticing things that other people, namely adults, seemed not to only deepened my sense of separateness. In spring I watched the hard beaks of purple crocuses crack the cold earth like hatched chicks. A red-bellied woodpecker drilled the enormous white oak outside my bedroom window. Each time I caught a creature in an act of unrestrained creatureliness, I felt I'd stolen a glimpse behind the curtain, into their world. The real world.

The best part of my childhood home was a midsize depression in the land some hundred yards into the woods behind our house. Each spring some two or three feet of rainwater swelled there and stayed nearly all year, icing over in December. In summer I would check my rubber boots for spiders and then amble out ankle-deep, petting the spongy mosses that clung to the tops of half-submerged rocks and greeting the skunk cabbages as if they were my friends. And they were, in a way. A mated pair of mallard ducks also lived in that swamp, but I didn't talk to them. They seemed to be socially occupied already; they had each other. The plants, on the other hand, seemed to have nothing else to do.

It wasn't so much that I imagined these plants to be little humans in another form. I don't remember ever thinking they talked back to me. But I also felt they weren't mute, exactly. They had their own thing going on. Like me. They were like children: underestimated.

In *The Ecology of Imagination in Childhood*, the writer and researcher Edith Cobb spends two decades investigating the role of nature in the early thinking of children. She finds that children have an "open-system attitude" that allows them a certain emotional proximity to the natural world. "For the young child, the eternal questioning of the nature of the real is largely a wordless dialectic between self and world," she writes. She references many artists and thinkers who describe their creative methodology to be essentially a channeling of the perspective they had as children; Bernard Berenson, a twentieth-century giant of art criticism, writes in his autobiography that his happiest moment, perhaps, was as a young boy, standing on a tree stump:

It was a morning in early summer. A silver haze shimmered and trembled over the lime trees. The air was laden with their fragrance. The temperature was like a caress. I remember—I need not recall—that I climbed up a tree stump and felt suddenly immersed in Itness. I did not call it by that name. I had no need for words. It and I were one.

Who doesn't have a memory like this? The "Itness" here bears so much resemblance to the sense of "the real" echoed by Sacks and Dillard and Von Humboldt. And to what I felt as a child, crouched and watching the crocuses. I wonder what moments like this *are*, and what they can do. What space they open up for thought.

Decades after leaving that house in the woods, I was a city dweller hermetically sealed into an office building. My nine-year-old sense of knowingness, of a world beyond the people-theater, was dulled to a slick nub. But then the fern obsession came over me, and then the plant intelligence debate. Something familiar began to tick quietly inside me again.

Lunchtime botany reading became my reason for the day. What I found in the scientific journals was some of the most acerbic controversy I'd come across in my years as a reporter. Just as common as the papers exploring plant intelligence were the responses denouncing the burgeoning field, most often for word choice. *Intelligence*, applied to plants, did not sit well with plenty of plant scientists. *Consciousness*, a yet bolder conjecture, even less so. They made good points; plants don't have brains, much less neurons. And plants evolved to meet challenges so different from our own. What need would they have for either of those things? A paper in *Trends in Plant Science* titled "Plants Neither Possess nor Require Consciousness," coauthored by eight highly credentialed plant scientists, appeared to tip off a series of heated back-and-forths. The authors wrote that it was "extremely unlikely that plants, lacking any anatomical structures remotely comparable to the complexity of the threshold brain, possess consciousness." Rather, they said, anything a plant did could be chalked up to "innate programming" via "genetic information that has been acquired through natural selection and which is fundamentally different from cognition or knowing, at least as these terms are widely understood."

The authors conceded that "excellent papers have been published" by plant consciousness proponents that make no truly contentious claims—even the ones about the role of electrical signaling in plant bodies, which, they concede, can be analogous (but, they are careful to say, not homologous) to animal nervous systems. The controversy, they wrote, stemmed from researchers who took their conclusions too far, who "risibly" oversimplified the meaning of terms like *learning* or *feeling* in service of the plausibility of their claims. "Why is anthropomorphism resurgent in biology today?" they lamented.

Science is a conservative institution for a reason. Conservatism is a crucial backstop against false knowledge. But something in this paper felt self-defeating. Science indeed has no agreed-upon definition for life, death, intelligence, nor consciousness. Words certainly matter, but the definitions of these words are not settled, and are therefore expansive. Could plants not hold intelligences that look quite different from our own? And the truth was, whatever electrical signaling pseudo-nervous system they were talking about sounded extremely compelling.

Science, for all its strengths, is constrained to the sort of questions that can be answered using the scientific method. The meaning or definition of life is arguably not one of them. Left to the sciences, which were never built to take up ethical questions of being and nonbeing, plants remain conceptually locked out in the inanimate cold. Yet here were all these scientists valiantly trying to engage with the hardest question of all, the very nature of being alert to the world; the hard problem of consciousness. And they were, after all, the stewards of the scientific information that might be used to come to any ethical conclusion about where plants fit, and how we might relate to them. Permitting or not permitting certain experiments to proceed, and be published, was entirely in their hands. I wanted to listen closer.

It was clear that the anti-plant-intelligence camp wished to be explicit that plants are not like animals. But they were using a human-centric definition of intelligence and consciousness to claim that plants couldn't possibly possess either thing. That argument seemed to me like it was marred by an internal contradiction; it doubled back on itself. Paco Calvo, a philosopher of cognitive science at the University of Murcia, and Anthony Trewavas, a well-known veteran of plant physiology at the University of Edinburgh, agreed: "This is surely circular reasoning."

Beyond that, I wondered if there was fear. I could see why those arguing against the idea of plant intelligence didn't want the narrative to slip away from them and enter mainstream culture prematurely, where it might be washed of complexity and absorbed in some diluted, fanciful form. Perhaps it would be used to support the same New Age notions that got them into so much trouble last time, with *The Secret Life of Plants*. This I understood, up to a point. There has always been an extreme tendency in popular culture to layer simple human narratives on other species, as seen in virtually every fairy tale or animated film. Still, it felt like a clear underestimation of that same capacity; the public's imagination was expansive, I thought. It could very well expand to include nonhuman types of intelligence, if given the chance. Yes, it was a tall order. Making the mental space to imagine truly different intelligences, without jumping to easy human conclusions, is a difficult task. Most of us haven't been asked to do that before. But grappling with complexity is a mind-expanding exercise. To hold back the larger scientific inquiry based on a fear about how it would be received seemed unfair to the rest of us. The world we could have if complexity was not backgrounded was the world I wanted to live in.

It seemed I had come to the plant intelligence debate early in its formation, but at just the right time. There were so many threads still to be picked up. Yet there was real science behind it, and the results coming in were too alluring to leave aside. What was at stake? Over and over, I saw the debate framed as a dispute over syntax. But it looked to me more of a dispute over worldview. Over the nature of reality. Over what plants were, particularly in contrast to ourselves.

THEY SAY TRYING to understand any culture is like looking at an iceberg: there are vast depths you cannot see. I saw the world of botanists and their culture—ideas they worked with and built on—as more like a rhizomatic plant. From where I sat, printing out and rifling through papers, I could see the shoots. The names, the concepts. But before long one botanist implored me to speak with another, who would in turn direct me to another. Networks of knowledge began to emerge, the many unseen underground connections between labs and journals. Who trusted who, and who didn't. Shoots and runners, shoots and runners.

Each time I called a scientist, I was reminded that most of them were utterly uninterested in yoking plants into human service. My best calls were with researchers totally in love with their subjects, the kind of love that you want to tell everyone about. Once convinced that I truly wished to know, they allowed themselves to unleash their volcanic enthusiasm. They told me about the corner of the world they themselves had just demystified, their own personal shard of the vast and chaotic puzzle of biology they had found by sifting with the finest sieve through the sediment of the world, turning it over in their hands, and through some mix of laborious years of reading and lab work and obsessive interest had known what meaning it held, where to set it in its place.

To see nature that way, I knew, was only a partial view. Nature is not a puzzle waiting to be put together, or a codex waiting to be deciphered. Nature is chaos in motion. Biological life is a spiraling diffusion of possibilities, fractal in its profusion. Every organism, and certainly every plant, has ricocheted out of another fragment of the evolutionary web of green leafy things to variate further. These each are of course still morphing, because that sort of thing never ends, except in extinction. The multiplicity felt endless and impossible to grasp. The scientists I spoke with knew this, and were trying anyway. It made me love them more.

I began to learn what to say—or, more accurately, not say—to keep a scientist on the phone. "Plant sensing" was generally fine, neutral territory. "Plant behavior" crept into more risky territory, and "plant intelligence" could be outright dangerous. Consciousness, I learned, I must only bring up once I'd made it through the crucible of each of these prior descriptors without feeling I might be hung up on or scolded. When I hit on a trigger word, I sensed it immediately. The researcher became cautious, less open, especially if we were still in the dance of sussing each other out, when I knew they were still deciding if they should be talking to me at all.

But every so often I'd feel the soft spot. Where there was give, where clearly they had their own curiosities about what behavior could mean, or what might count as intelligence. They would thoughtfully consider my questions and, after some hesitation, give thoughtful answers. Often this revealed inner conflicts. To many people I spoke to, "intelligence" felt dangerous, yes, but only because most people make a mental leap directly to human intelligence. Measuring plants against human cognition made no sense; it just rendered plants as lesser humans, lesser animals. Anthropomorphizing was dangerous because it diminished these green bodies, leaving no room for the recognition that plants deploy several senses—or could one say, intelligences?—that far exceed anything humans can do in a similar category. Our versions of those senses, if we even have them, are paltry in comparison. It was hard for these researchers to talk about plant intelligence; they worried that it was a trap, leading to conclusions that did not represent the real wonder of what they'd learned.

BY NOW IT had been more than a year since I'd first fallen into these questions. It was August 2019 in New York City, and the air was thick with the tang of heated garbage and pavement. Every day I exited my sweltering Flatbush apartment and walked the six blocks to Prospect Park. I sometimes stopped to buy a cold water coconut (a seed) or a few inches of sugar cane (a grass) from a man who sold them off a card table at the corner. When I stepped past the park's stone columns, I slowed down. The light darkened and the temperature dropped, thanks to the exhale of millions of plants at once. I remembered that prior to the discovery of photosynthesis as a reaction to produce sugar, naturalists believed that its purpose was to serve as nature's air conditioner. The cool air settled on my skin, and I inhaled fully. It smelled like wet leaves, clean and head-clearing. I now regarded the broadleaf plantain and chokeberry with a mix of awe and suspicion. I was newly aware that there was more going on in the lives of each plant I passed, both aboveground and below, than I'd ever imagined. That they might well know I was walking by. Amid the wash of greens, bright and dingy, I began to see many distinct species, and exponentially more individuals. I knew there was high drama everywhere I looked, though I couldn't see it, nor fully know the nature of it.

I was regaining material intimacy with the natural world by looking at plants. It wasn't a way to ignore environmental catastrophe; it was a way to reattach myself to the stakes. Each plant was an embodied world we stood to lose, every ecosystem another galaxy. Still, reading plant intelligence papers felt like trying to understand a mountain by looking only through a hand lens. The flurry of recent discoveries merely underlined that. Researchers had just found that plants could remember, but not where those memories were stored. They'd found kin recognition, but not how those kin are recognized. These discoveries were more like hints, fragments that pointed toward something larger, something whole.

What was a plant? No one yet seemed to know. I decided, on that walk, to quit my job, to think about plants full-time. The newsroom where I worked had fallen on hard times. People were laid off as ad revenue dipped, and investors got spooked. Morale sank through the floor. I could no longer see the point in my being there. I felt I could be dismissed at any time, so that even the security of full-time employment began to seem like a lie. I had some savings, I would downsize my life. It was time for a change. A childhood friend had room for me in the old farmhouse on the farm where he'd grown up, where we had run through rye fields as kids. I could situate myself there, and then travel to see plants in more places, in their ancestral habitats, the places they'd evolved to live in.

It would be worth it; something clearly important was happening in botany. Science was approaching a precipice from which it couldn't possibly return: our belief that plants are mute, unfeeling beings seemed to be wholly wrong. The moment felt ripe. It was a good story, too good to stay locked in the obscure realms of academia. I began to feel like it could change the world. Certainly it was already changing mine. The stakes of the story began to feel much bigger than my own personal attraction to it. Or perhaps, I thought, they were one and the same. The more time I spent thinking about plants, the more I wanted to spend time thinking about plants; it felt abundantly good for me. I felt I could see everything more clearly.

I walked back home and looked up at the gigantic golden pothos hanging in my kitchen window. All of its leaves were upright. They had all swiveled to face the windowpane since I'd been gone, practically plastering themselves to it. I glanced around at my other houseplants. The philodendron was now poking a slender brown aerial root into the potted jade beside it. I looked at my rubber plant, which was a cutting from my father's rubber plant, which was itself a cutting from his parents' rubber plant, which had been a gift on their wedding day sixty years before. That original plant, now a formidable tree, still sat by the grand piano in their living room, lording over the scene. It once nearly died; my grandmother's mother lopped off its surviving branch, soaked its blunt end in water until white roots slithered out, and coaxed its complete regrowth from that single healthy limb. Four generations of my family had handled this plant, and here it still was, silently building new body parts. Wasn't this a kind of memory in itself?

It had become unbearable not to understand. I had to go out and see for myself.

Chapter 2

———

How
Science Changes
Its Mind

Facts are theory-laden; theories are value-laden; values are history-laden.

—DONNA HARAWAY, *IN THE BEGINNING WAS THE WORD: THE GENESIS OF BIOLOGICAL THEORY*, 1981

To ask humankind what being in the world means . . . is to reproduce a very partial image of the cosmos.

—EMANUELE COCCIA, *THE LIFE OF PLANTS*, 2019

The roiling plasma surface of the sun flings out a fistful of light. The particles—billions of photons—hurtle 93 million miles through black space to rain down like bread and honey on the outstretched flesh of the most abundant living mass on earth. Plants eat light. Photosynthesis, so basic to plants, is the prerequisite for most every other life form on earth. Through photosynthesis, plants suffuse the air with the oxygen we breathe.

How did we get here? A billion and a half years ago, an alga-like cell swallowed a cyanobacteria. That alga-like cell was the early organism from which both animals and fungi would later evolve, and the cyanobacteria is an ancestor of the unthinkably diverse bacteria that flood our world today. But together they were the start of a new branch of life entirely.* Afloat in the murky waters of the Precambrian, this single sentinel of a new kingdom began to photosynthesize. It took sunlight, and alchemized the spare materials of its environment—water, carbon dioxide, maybe a few trace minerals—into sugar.

The first plant was born a chimera, an organism composed of cells that are genetically distinct. The leaves of every green plant on Earth retain the genetic imprint of that first union. The plant cells that today catch photons as they fall from space are themselves chimeras in miniature; that first cyanobacteria is still within them, still faithfully alchemizing light into food.

A billion and a half years after that first creation, plants have evolved and proliferated into a half million species that thrive in every ecosystem on the planet. Their supremacy is absolute. If weighed, plants would amount to 80 percent of Earth's living matter.

When plants climbed out of the ocean some five hundred million years ago, they arrived in a terrestrial barrens enveloped in an inhospitable fog of carbon dioxide and hydrogen. Inhospitable, that is, to everything but plants. They had already learned to unlock oxygen from the carbon dioxide dissolved in the ocean. They adapted the technology to their new world. In a way, they brought the ocean up with them. By incessantly breathing out, those legions of early land plants tipped the balance of gases toward oxygenation. They created the atmosphere we now enjoy. It's not a stretch to say they birthed the habitable world.

---

* A third organism, a bacterial parasite, was also involved—it was the go-between, ferrying the food from the domesticated cyanobacteria to the host alga-like cell.

As the Italian philosopher Emanuele Coccia puts it, they constructed our cosmos; "The world is, above all, everything the plants could make of it."

Through the same process, plants have made every iota of sugar we have ever consumed. A leaf is the only thing in our known world that can manufacture sugar out of materials—light and air—that have never been alive. All the rest of us are secondary users, recycling the stuff the plant has made. Our recombinations may be genius, but the matter is not original. The original is made like this: As photons from the sun fall upon a plant's outstretched green parts, chloroplasts in the leaf cell convert the particle of light into chemical energy. This solar power gets stored inside specialized energy-storing molecules, the rechargeable battery packs of the plant world.

At the same time, the leaf siphons carbon dioxide out of the air through minuscule pore-like openings on the underside of the leaf, called stomata. Under a microscope, stomata look like small parted mouths, fish lips that open and close. They are breathing, after all, in their way. The stomata suck in carbon dioxide, and the carbon dioxide now encounters both the stored solar energy in the chloroplast and the water that is coursing, always, through the leaf's veins. Through that encounter with the pure energy of light, the water and carbon dioxide molecules are ripped apart. Half of the oxygen molecules from both parties float away from this meeting, passing back out into the world through the parted lips of the stomata—becoming the air we breathe. The carbon, hydrogen, and oxygen that remains is spun into strands of sugary glucose. To be precise, it takes six molecules of carbon dioxide and six molecules of water, torn apart by power from the sun, to form six molecules of oxygen and—the true aim of this whole process—one precious molecule of glucose. The plant uses the glucose to build new leaves, which will be used to make more glucose. It also shuttles the glucose down through its body, passing it into its underground architecture, where it is used to grow more roots, which will pull more water back up through its body, which will be torn apart to make more glucose. In this way, life unfurls.

We are made of glucose, too. Without a constant supply of the plant sugar, our vital functions would quickly cease. Think about it: every animal organ was built with sugar from plants. The meat of our bones and indeed the bones themselves carry the signature of their molecules. Our bodies are fabricated with the threads of material plants first spun. Likewise, every thought that has ever passed through your brain was made possible by plants.

This is crushingly literal. The brain in particular is a machine run chiefly on glucose. Without a continual source of glucose, communication between neurons will slow and then cease. Memory, learning, and thinking will shut down. Without glucose your brain will wither, shortly before you do. All the glucose in the world, whether it arrives in your body packaged inside a banana or a slice of wheat bread, was manufactured out of thin air by a plant in the moment after photons from the sun fell upon it.

In this way we are, at every moment, brought into conversation with plants, and they with us. Our thoughts, and the products of them—the fabric of our cultures, the direction of our invention—have behind them trillions of plant bodies, each alchemizing the world into existence.

Yet for all they can do, plants cannot run around. It is perhaps one of the greatest feats of life that plants were able to distribute so broadly, given their limited mobility. Colonizing all seven terrestrial continents required innovation, adaption, and luck. But arriving was only one feat. To survive, reproduce, and establish complex communities—all while thwarting the pressures of predators, seasons, scarcity, and blight—was another thing entirely.

NO ONE KNOWS this better than a rare-plant botanist working on a far-flung island. Steve Perlman is the top botanist at Hawai'i's Plant Extinction Prevention Program. When I meet him, he is sixty-nine, with a sturdy build and white hair. Before I got into the thorny world of plant intelligence research, I wanted to see some straightforward, old-school botanizing. I've arrived to see his work, but for now, as we bump along in an old minivan as it struggles up a winding clay road on the north-western edge of the island of Kaua'i, we talk about feelings. He doesn't take Prozac, like some of the other rare-plant botanists he knows. Instead, he writes poetry. Either way, Perlman tells me, you have to do something when a plant you've long known goes extinct. Each plant that dies that singularly lonely death marks the end of a multimillion-year evolutionary project. That species' great genetic experiment is over; it's the last in its line.

Every native plant on Kaua'i, Hawai'i's fourth-largest island and Perlman's home base, is a mind-blowing stroke of luck and chance. Each species there arrived on the island as a single seed floating at sea or flying in a bird's belly from thousands of miles away—more than two thousand miles of open ocean sit between Kaua'i and the nearest continent. Botanists believe one or two seeds made it every thousand years.

Kaua'i was formed by a volcano five million years ago, and was pushed off the volcanic hotspot by plate tectonics. The island is still drifting northwest, little by little, every year. Another island, then another, would emerge at this geologic birth site and drift leftward the same way. Because it was the first Hawaiian island, and is therefore the oldest, it has had the most time to collect errant seeds. When a new seed took root in its young soil, the plant would evolve into a completely new species, or more often several new species, each trying out a different lifestyle in the comfortable embrace of the island's perfect climatic conditions. This process is known as adaptive radiation. The result is thousands of variations on a few species; each new variation became endemic (found exclusively on the island).

I try to hold the grandeur of that fact in mind while I look out the window of the bouncing minivan. Perlman is driving. Bushy fronds sweep the side of the van like gloved hands.

A precipice on one side of the road plunges several thousand feet down, opening into a canyon covered in pale greens. The higher we ascend, the thicker the fog enveloping the van grows. Soon the dense vegetation outside the window has become a wet, green smudge. The road plateaus, and Perlman parks and walks out. We are very high up now.

He strides out until the tips of his work boots hang over the edge of the cliff and looks down. The vertical drop covered in ferns resembles a shaggy fur coat, with small palms sticking out at odd angles, poking through the mist. The cliffs form a small half-moon valley at their base, its other edge the Pacific Ocean. It is green in every shade for thousands of feet down. Pearlescent wetness clings to everything like spider's silk.

In many ways, Kaua'i is the ultimate example of what a world would look like if plants were in charge. The whole island is covered in the surreal products of total floral freedom. When plants are allowed to evolve without fear, they get scrupulously and flamboyantly specific. Take the *Hibiscadelphus* genus, for example. Found only in Hawai'i, these plants have long tubular flowers, custom-made to fit the hooked beak of the honeycreeper, the precise bird that pollinates them. Then there is the vulcan palm, *Brighamia insignis*, or 'Ōlulu in Hawaiian, a short tree best described by its nickname, "cabbage on a stick." Over tens of thousands of years, it has evolved to be pollinated only by the extremely rare fabulous green sphinx moth (its real name).

The vulcan palm, still critically endangered in the wild, was saved from total extinction by Perlman's work in the early days of the extinction prevention program, when he made his own harness out of knotted ropes and used it to hang over the Nā Pali Coast cliffs. There, four thousand feet in the air, he would use a small cosmetic brush borrowed from his wife to imitate the moth, carefully transferring the pollen from the males to the females. "You'd know if you did it well," Perlman said. "When you'd go back, there'd be fruits just bursting open with seed." (The vulcan palm is now cultivated as a houseplant in the Netherlands, where there are greenhouses full of them. I wonder if a person with a potted Vulcan palm on their Amsterdam windowsill knows of the drama it took to get it there.) Other plants adapted to live at very specific heights, where, say, the mist that drips from ferns clinging to the cliff just above them creates a perfect moisture balance.

Outside of Kaua'i, in most every other place on Earth, plants have had a very different evolutionary trajectory. The first plants with seeds and flowers appeared some two hundred million years ago. Since then, they have split and evolved into hundreds of thousands of species that have had to adapt to threats of all kinds, which start the moment they sprout.

When a seed decides to take root, it makes a tremendous gamble. Seeds are embryos encased in nutrients; a seed scientist once described them to me as a "plant in a box with its lunch." The blueprint for the whole plant is in there, dormant but alive the whole time. A seed might blow around for a decade, waiting patiently for the right conditions to poke out a first root. Once it sends this first root out, it has given up any chance of movement; immobilized, it will now face whatever threat comes—wind, snow, drought, animal mouths—from right where it stands.

The baby plant root has forty-eight hours after it decides to emerge to locate water and nutrients, and then push out a leaf or two and begin photosynthesizing, before it runs out of resources and dies. The first green parts of any plant are folded, preassembled and waiting inside the seed. This preassembled plantlet bears little resemblance to the plant itself; it consists of one or two cartoonish green lobes on a short green stem, the manifestation of the plant emoji, and it is entirely temporary. It unfolds and inflates, filling with the first draught of sap pulled in by that pioneer root, and begins the job of photosynthesis. If it is successful, this proto-plant, this space shuttle into the world of air and light, will be cast off like a rocket booster and replaced with real leaves, of which the variations are infinite. Only after this trial period, this checking to see what sticks, does the plant come to resemble the one it was meant to be, mantling itself in the features of its lineage and then adapting them to a new environment.

Even then the plant has overcome only the first of many threats to its young life. Any given seed has a vanishingly small chance of making it into its full plant form. Many of these threats are grazing animals; creatures that can run and forage across a broad terrain, and whose essential functions include a central nervous system. Plants have none of these advantages. They cannot run away; instead, they've developed ingenious and complex means of defending themselves from their tormentors, as well as ways to eke out a lifetime of nutrients from the very spot they first landed as seeds.

The dangers of being immobile are precisely the forces that have driven plants to engineer some of the most impressive adaptations in nature. Perhaps a plant's biggest achievement is its anatomical decentralization. A plant is modular; snap off a leaf, and it can grow a new one. Without a central nervous system to protect, the plants' vital organs are distributed and come in duplicates. That also means a plant has evolved remarkable ways to coordinate its body and defend itself. They might grow thorns and spikes and stinging hairs, developed with remarkable precision, to pierce the flesh or exoskeleton of whatever mammal or bug might be its main threat. They might secrete sticky sugar to entice and then immobilize their antagonists, whose hungry mouths get stuck shut. Their flowers might be extra slippery, to deter nectar-thieving ants. Whatever the adaptation, it tends to be economical in its specificity. There's a purpose to every tiny variation. This is true for all areas of plant physiology; every part of the architecture of a plant's body is there for a reason, calibrated to fit its task. No more, no less.

Any sense that immobility implies passivity is quickly banished by a look at plants' vast capacity to make chemical weapons. Plants are themselves synthetic chemists, surpassing the best human technology in terms of the subtle complexity of the chemicals they can synthesize. A leaf, sensing that it has been nibbled, can produce a plume of airborne chemicals that tell a plant's more distant branches to activate their immune systems, manufacturing yet more repellent chemicals to deter incoming aphids and other plant-eating bugs. Several species of plants have been found to identify a caterpillar's species by sensing the compounds in its saliva, and then synthesize the exact compounds to summon its predator. Parasitic wasps then obligingly arrive to take care of the caterpillars.

But Kaua'i's plants have none of these defenses, or in any case, far fewer. Any that the plants' predecessors may have had—thorns, or poison, or repellent scents—were completely dropped after they landed on the island. No large land mammals, reptiles, or other potential predators made the journey from the mainland to the remote island chain. In fact, the only land mammal native to all of Hawai'i is a small, fuzzy bat. (The journey its ancestor must have taken from North America is almost unthinkable; it likely gusted over inside a storm.) From the plants' evolutionary perspective, there is no reason to spend energy on defenses when there are no predators to fend off, so mint has lost their mint oil, and stinging nettles do not sting. Scientists ominously refer to this process as species becoming "naive."

Once threats show up, this blissful naïveté is often fatal; Kaua'i, like the rest of Hawai'i, is now beset by invasive species that evolved elsewhere, in less comfortable conditions. Those species are more aggressive—or resourceful, to use a less emotionally loaded term—because they have to be. In Kaua'i, they take over ecological niches easily. Native plants don't stand a chance. As a result, Hawai'i is losing plant species at the rate of one per year, compared to the natural background rate of roughly one every ten thousand years. That's where Perlman comes in. He and his field partner, Ken Wood, deal exclusively in plants with fifty or fewer individuals left—in many cases, much fewer, maybe two or three. Of the 238 species on that list at the time of my visit, 82 were on Kaua'i.

Without Perlman, rare Hawaiian plants die out forever. With him, they at least stand a chance.

To reach these plants, Perlman rappels down cliffs, and sometimes jumps out of helicopters, to reach a cluster of sometimes as few as five plants clinging to a remote cliffside on a Pacific island. In cases where the remaining male plants are too far from the last remaining females to naturally pollinate and reproduce, Perlman will bag the pollen from the males, lovingly carry it to the females, and apply it to their sexual parts with a paintbrush. Finding these plants involves trekking for days, eating granola bars and foil packets of tuna to save precious space in his pack for bulky botanizing tools. Sometimes he arrives at the plant too soon—it hasn't sexually matured, perhaps, and the flower hasn't yet opened—and the whole endeavor must be postponed.

The time-intensive nature of this process means that Perlman develops relationships with plants he aims to save. He isn't always successful; that would be impossible. "I've already witnessed about twenty species go extinct in the wild," he says. He has sat beside the last of a species and kept it company as it died. Plant death, like human death, is a question of both biology and philosophy. Is a person dead when their heart fails? Or when their brain does? A plant can technically be reproduced in a lab from just a few living cells. But a plant with a few living cells left is not a thriving plant. Perlman considers a plant dead when enough of its tissues have succumbed that it no longer has any chance of thriving in the wild. It dehydrates, wilts, turns brown, collapses.

To Perlman, that a plant's evolutionary invention would end there is enough of a reason to save it. You don't just abandon a species, not if you can still get to it, even if it's on the face of a remote and jagged cliff. Or, as Wood puts it, "We try, because we're not going to not try." Sadly, failure is part of the job. Once, when the last known individual of a native flower finally withered and died, Perlman dug up the plant and brought it to a bar. Overcome with emotion, he raised a toast to the plant's life.

It's fair to say that most people have little feeling toward rare plants, and fewer have any idea of the fight going on to bring them back from the brink. While the average person can differentiate between several types of canines, they are far less likely to know a beech from a birch, or an ear of wheat from an ear of rye. This is understandable; plants are much further from us evolutionarily, having evolved in a context so unlike our own. They make food out of light and grow rooted to one spot, spending decades or centuries probing their environments for sustenance. Their way of life is so alien as to often preclude them in our imagination from even having a way of life.

This state of unseeing has become a named affliction, lamented among botanists: "plant blindness," the tendency to view plant life as an indistinguishable mass, a green smudge, rather than as thousands of genetically separate and fragile individuals, as distinct from one another as a lion is from a trout. The term shows up in research papers and at conferences where worried scientists wring their hands over how to get the public to even see the subjects of their life's work. For botanists, plant blindness amounts to a perennial struggle to get basic research funded, or to convince anyone that a particular plant needs saving if it falls outside the few plants included in the human economy, like the type of corn with the highest starch content, which we use to feed cows, or the two species of coffee we drink.

Generally speaking, humankind knows very little about the tender green flesh that frames our every landscape and populates almost every inch of unpaved ground around us. The plant kingdom keeps its secrets well hidden from a species that doesn't bother to look. But they have the supreme authority to influence our biology and culture, a sphere of influence shared, it should be said, by bacteria and fungi, which are similarly ignored. We seem to have been stricken by poor judgment, our allegiances preposterously misplaced.

A COMMON EXPLANATION for our general lack of interest is that plants are slow. Their world exists on a different time scale than ours. It is true that we tend not to see their daily movements, like the way a young cucumber plant may curl and uncurl its tendrils and sway back and forth several times a day. It happens slowly enough that only the most patient among us could possibly notice. Still, slowness is relative. A forty-year-old tree will be much, much taller than a forty-year-old man. A bean plant can grow to the height of a ten-year-old child in less than a month. Kudzu can engulf a car in the space of two weeks.

It seems to me that plant blindness is something deeper, more tied to value systems, which are of course a product of cultural perspective. Indeed, not all cultures have this problem. Virtually all Indigenous groups around the world have a more intimate relationship with and recognition of plant life. Many cultures ascribe personhood to plants, humans being just one type of person. Human persons and plant persons are often literally related: the Canela, a group of Indigenous peoples in Brazil, include plants in their family structures. Gardeners are parents; beans and squash are their daughters and sons. In *Plants Have*

So Much to Give Us, All We Have to Do Is Ask, a collection of traditional Anishinaabe teachings about plants, Mary Siisip Geniusz writes that the primacy of plants is central to the understanding of her Great Lakes-area people. Plants are the world's "second brothers," created just after the "elder brother" forces of wind, rocks, rain, snow, and thunder. Plants are dependent on those elder brothers for their life, while supporting all life created after plants. Nonhuman animals are "third brothers," reliant both on the elements and on plants. Humanity is the "youngest brother," the most recently created of all the beings. Humans alone need all three of the other brothers to survive at all. "Humans are not the lords of this earth," Geniusz writes. "We are the babies of this family of ours. We are the weakest because we are the most dependent."

Where Geniusz talks of linkages, of dependance and kin, most European thought is fixated on distance and detachment. Perhaps nowhere is this more clearly illustrated than in our corruption of the word *vegetable*, which is now a crude word for a brain-dead human being. But *vegetabilis* came from the medieval Latin, meaning something that is growing or flourishing. *Vegetāre*, the verb, meant to animate or enliven. *Vegēre* was the very state of being alive, being active. Clearly, it hasn't always been this way.

I think of the theorist Jane Bennett, who is interested in the language we use to talk about the liveliness of nonhuman things. We are too seriously engaged in the funny business of drawing lines in the sand between subjects and objects, she says. "The philosophical project of naming where subjectivity begins and ends is too often bound up with fantasies of a human uniqueness," she writes in her book *Vibrant Matter*. Or else it relies on a belief in our supposed mastery of nature, or our superiority in the eyes of God, or some other thin claim, making the whole thing rather starry-eyed and materially useless. We can do better than that, I thought.

SO HOW DID the white European perspective on humanity's place in the world veer so far from the unequivocal reality of our own dependence on plants?

The roots of the answer run deep. In ancient Greek philosophy, almost as soon as a "soul" came to differentiate animate from inanimate things, plants were included among the soul-havers. Empedocles gave plants souls in his accounting of the world, and referred to them as animals, precisely because they were animate—alive—and he felt no reason to cleave that category. Later, Plato described plants as having a "desiring" and "sensing" soul, which, albeit the lowest of the various souls, was also endowed with intelligence, simply because, to Plato, there was no sensation without intelligence, and no desire without intent. Humans had these desiring-sensing souls too, but those were improved greatly by rationality and morality, which made humans—particularly free men—a special case. Rationality was becoming the mark of superior sense, something Plato believed only men could grasp, and women, children, and slaves typically could not. The rational thing, then, was for men to rule over these lesser people, as well as all of nature.

Writing a few years later, Aristotle intensified that hierarchy. He described a ladder of life, a *scala naturae*, with plants on the bottom and humanity at the top. There is no intelligence there at the bottom, he argued, nor even sensation. One rung up, animals have sensation, but no rationality. By this time, Greek philosophy was shifting overtly toward a fervent belief in rational cause and effect, and turning away from the Ancient Greeks' beliefs about the need to maintain respectful relationships with other living things. Keeping the peace no longer required ritual acts of deference to the elements and nonhuman creatures, but rather simply a rational understanding of what caused natural phenomena. Aristotle stripped plants of even the earlier ability to desire or sense; they existed wholly as instruments of man.

Here we come to a fork in the road. It surprised me when I first came across it. This fork was named Theophrastus, and he presents an alternate ending to this story, a path Western thought did not take, but could have. When Aristotle died, he left his school to Theophrastus, a standout among his students. Theophrastus took a special interest in plants. He published the first known texts about plants themselves, as opposed to merely what purpose they may serve humans. He described their behavior: how they grew, what they pursued, and their apparent likes and dislikes. Plants, he wrote, were not at all passive but rather in constant motion, seeking after their desires. Incredibly, he described agriculture as a collaborative relationship. He saw that cultivated plants appeared to suffer from a shorter lifespan than their wild counterparts, but he thought that plants might weigh their shorter lifespans as a reasonable trade-off for the many benefits of being protected from predators and given all the food and water they need.* Theophrastus appeared completely willing to take the plant seriously as an autonomous being with desires and the will to satisfy them.

Just as intriguingly, Theophrastus articulated the ways plants were entirely different from animals and humans, without passing any judgment on where that placed them on an imagined hierarchy, as Aristotle had done. He did draw certain parallels between people and plants—he notably equated the liquid flowing through plants to the blood of animals, remarking that both flow through veins, and described the core of trees as "heartwood," a term we still use today. But he was quick to clarify that he did not think of plants as simply inferior humanoids. They were entirely their own category of being, not to be measured against animals. The comparisons between hearts and cores were useful merely as a bridge to aid understanding. "It is by the help of the better known that we must pursue the unknown, and better known are the things larger and plainer to our senses," he wrote. This seems to me profoundly humane. He was meeting his readers where they were, using metaphors that spoke to them, blindered as they were by their human perspective. In short, in writing about the complexity of plants, he recognized the limits of people. What would modern history look like had Theophrastus's model prevailed?

But by some trick of time and prevailing fashions, it was Aristotle's

---

* The plant may "accept these internal modifications as appropriate to itself; and it is reasonable that it should demand and seek them," Theophrastus wrote. *De Causis Planatarum* 1.16.12.

hierarchy, not Theophrastus's, that clung to the natural sciences, and Western morality, ever since. The result? Lots of things. But perhaps none were more symbolic of this inheritance than the surgical dissection of fully awake dogs in amphitheaters until nearly the dawn of the twentieth century.

Aristotle believed humans had "rational souls," but all other animals had only "locomotive souls," propelling them forward, without thought, toward reproduction and survival. That general idea held sway in the Western world for two millennia, and was renewed in the seventeenth century by French philosopher and scientist René Descartes, who believed animal bodies were just solvable puzzles of physics and chemistry, popularizing the notion of the "animal machine."

The idea was that "vital phenomena, like all other phenomena of the physical world, are capable of mechanical explanation," as biologist Thomas Huxley put it two hundred years later, in 1874. The passage of time had only reinforced a Descartian grip on the sciences, because each new advancement of the era seemed to support it. Physiology and anatomy had made crucial discoveries about the way bodies worked how we digest our food, breathe, and move. Each proved itself quite mechanistic. European men of science felt they were on the verge of discovering the life force itself, which would surely prove to be just another mechanical ingredient, like blood or bone. This was the era of Frankenstein's monster: if you put the pieces together correctly, you just might make raw life itself.

But humans, despite the mechanistic nature of their bodies, had an ineffable sense of reason and a soul that distinguished them from other animals. Dogs, it was then thought, did not. The way a dog perceives its environment, or even feels sensation, were not truly conscious experiences but rather the rote reflexes of an automaton. Any expression of pain, like barking, was the same; just a reflex. This was all considered scientific fact. And the mechanical basis of animals absolved humans of any guilt when they dissected them alive for scientific study.

In the 1800s vivisection, as it was called, came back en vogue, and led to new scientific understanding. The English physiologist William

Harvey was the first European to accurately describe how blood circulated, thanks to dissecting live animals. (Ibn al-Nafis, an Arab physician from Damascus, beat him by a long shot, accurately describing pulmonary circulation three hundred years prior.) Claude Bernard, the renowned French physiologist, supposedly live-dissected his family dog in the 1860s. The story goes that after coming home to find what Bernard had done, his wife and daughters left him to join an early antivivisection society. Vivisection fell out of fashion not because science had changed its mind, but because the first animal welfare societies—led in most cases by women—sprang up to oppose it.

Reflecting on the way animals were up until very recently viewed is useful to our story about plants because it serves as a potent example of the fluctuations of scientific opinion. It also shows how philosophy and ethics can come to intervene in the way non-human creatures are viewed. If left entirely up to science, it would likely have taken much longer (if it came to pass at all) to view animals as worthy of some semblance of humane treatment. We don't think much, now, of the fact that we grant at least some animals the benefits of personality and intelligence. We've also decided that it is cruel to do them harm. Of course, the morality surrounding what one should or shouldn't do to animals is still quite broadly permissive, and we play favorites with species. But the point is only that an ethics of humane treatment exists now that didn't before, and we take it utterly for granted.

In fact, scientists ascribing consciousness to any animal is so recent that the internet is older. In 1976 a zoologist named Donald Griffin published *The Question of Animal Awareness*, a book arguing that animal cognition ought to be taken seriously. He and a colleague had been responsible in 1944 for discovering that bats navigate by echolocation. Now, after a lifetime spent watching those creatures, he became convinced they had inner worlds. They had flexible behavior, he said, or the ability to change their behavior as external circumstances changed, a hallmark of true intelligence. He'd watched bats develop ingenious techniques for finding food; they could clearly make decisions on the fly, and exhibited many of the same problem-solving abilities as humans did. Animal thought and reason ought to be legitimately studied, he argued. After all, despite the flourishing of neuroscience, no one had yet found any part of the brain unique to humans that might impart this hallowed "consciousness." Wasn't it time to give up the ghost?

Griffin was widely criticized for the sin of anthropomorphization. It was years before any of it was taken seriously. But his writing put the idea of animal consciousness on the map.

It took the neuroscience revolution of the 1960s for researchers to think of the "mind" as something scientists could study by watching people's behavior, rather than by directly observing their brains. By the 1990s and 2000s, ambitious zoologists were using those techniques on dolphins, parrots, and dogs. They found that elephants could recognize themselves in the mirror, that crows make tools, and that cats exhibit the same attachment styles as human toddlers.

Today, just four decades since Griffin's plea to his field, it isn't heresy to talk about animal cognition, to study the behaviors of individual animals, and to ascribe to them personalities. In fact, it's approaching the mainstream. In 2012, a group of scientists gathered at the University of Cambridge to formally confer consciousness on all mammals, birds, and "many other creatures, including octopuses." Nonhuman animals had all the physical markers of conscious states, and clearly acted with a sense of intention. "Consequently, the weight of evidence indicates that humans are not unique in possessing the neurological substrates that generate consciousness," they declared.

It was a short list; mammals, birds, octopuses. Yet everywhere researchers look, it seems, there is much more to the inner lives of all animals than we ever thought possible. What comes after mammals and birds, in our conception of the order of species? Maybe reptiles, maybe insects? Lizards have been shown to learn how to navigate mazes, suggesting behavioral flexibility, that often-used marker of intelligence. Honeybees were recently found to be able to distinguish between styles of art, and individual bees perform an elaborate, symbolically rich "waggle dance" that tells their hive mates precisely how far and at what angle to the sun to fly to find food. New research suggests bees may have a form of subjectivity, a marker used by some to denote consciousness. Where further down to look, past insects? How about plants?

Right now, one camp of botanists is arguing that it's absolutely time to expand our notions of consciousness and intelligence to include plants, while another argues that's an illogical road to go down. Many more botanists are sitting in the middle, quietly doing remarkable work, waiting to see how the larger debate falls out. I've been sitting with them. I don't know where it will fall either. But I do believe we are standing at the precipice of a new understanding of plant life. Science can feel like a monolith; what it says to be true now will always be true. But things can change fast.

ONCE, AT A rare botanical book library in rural Virginia, I got to touch the products of a different time in botanical understanding. Botanical texts, hand-painted on handmade paper, were once the pinnacle of botanical technology. They told readers how to heal themselves with plant poultices, or gave them a first glimpse of the foliage of a faraway continent. Very often they simply broadcast status; these were luxury objects, the product of hundreds of hours of a person's toiling with fickle pigments on delicate paper by lamplight. The library in Virginia was the private collection of a wealthy philanthropist, now deceased. It was open by appointment only, and few people outside the rare book world seemed to know about it. As such it had a tranquil air of a secret garden, with blond wood shelves stretching to the ceiling and thousands of volumes from the fifteenth century to the nineteenth. Tony, the knowledgeable librarian, acted as a book sommelier, assessing your interests and pulling out volumes he thought you might like.

Looking at old botanical texts is a pure treat—the colors, the heft of the old handmade paper, the incredible attention to minutiae that make plants look as lively as a panther, ready to jump off the page. But I get the most pleasure from old books that were clearly passion projects instead of commissions. They don't broadcast status; sometimes the plants are perfectly ordinary to their region, with very few fancy imports. The illustrations may even be a little juvenile; the daffodil may have a slightly lumpy look, or a crocus have too thick a line for a stem, breaking the visual fourth wall. These are personal books, sometimes kept and added to throughout the painter's life. Tony thought I might like to see one of these. He pulled a thick volume from the shelf, bound in leather. It was a personal chronicle without words. Charles Germain de Saint-Aubin began the book as a teenager in 1721, painting one page at a time, and adding to it until he died in 1786. His style and refinement changes over the course of his life, technically improving, but to me that seemed hardly the point. All of the images had the same emotional quality—the devotion to these plants, the tender need to record them at their peak bloom and put them in little arrangements. Making their forms permanent through painting seemed an obvious pleasure; these were portraits of the flowers and fronds with whom Saint-Aubin shared a life. Plants were clearly more than decorative. They seemed almost like companions.

Botany, the study of plant life, is as old as human thought. Yet questions about plants' life, how they actually lived, took longer to show up in the literature. The mysteries plants hold have always been vital to survival, and as such, information about plants in their capacity as food and medicine shows up in the earliest examples of writing, based no doubt on knowledge passed orally for thousands of years before that. Prior to the advent of pharmaceuticals, plants and fungi—a kingdom of life all their own, and frequent plant-collaborators—were the medicines for anything that ailed us.

The first written information about plants themselves, rather than their uses to humans, appeared in Theophrastus's *Historia Plantarum* around 350 BC. It classified plants into categories based on their structure, reproduction, and growth. It's often regarded as the first text of plant science. But it would take more than two thousand years for plant behavior to finally enter Western literature. By the late Victorian period, botanizing had become a popular pursuit of the wealthy intellectual class, which was operating under the same assumption as always, that plants were utterly inactive, rocks that happened to grow. Its focus was almost exclusively on classification and botanical illustration.

Then, in the 1860s, Charles Darwin became captivated by plants. By then he was already a famous man. It had been a number of years since he'd published *On the Origin of Species*, and island voyages, exotic animals, and volcanic geology seemed to better suit a younger him. As an older man, he shifted his focus to nearer things, right at his feet: almost all of his books after *Origin* were about plants. As such, we'll hear about him over and over in this book.

Over the course of dozens of experiments that led to several books, Darwin observed how plants moved with remarkable athleticism, albeit very slowly (*On the Movements and Habits of Climbing Plants*, 1865), how they sometimes produced curious, irregular versions of themselves (*The Variation of Animals and Plants Under Domestication*, 1868, and *The Different Forms of Flowers on Plants of the Same Species*, 1877), and how carnivorous plants used tricks to lure and eat insects (*Insectivorous Plants*, 1875). He was treating plants as subjects that had activity and purpose.

The Power of Movement in Plants, his second to last publication, was an inquiry into why plants moved the way they did. It was filled with experiments on plant roots, done alongside his son Francis. The conclusion they came to was startling. The very end of a plant's root, he wrote, is covered with an unassuming cuticle that seems to be a command center. Poke it, or singe it, and the root will grow away from the offending jab. Place moist and dry soil on either side of it, and it will swerve toward the moisture. Put it between a rock and softer clay, and it will, every time, move away from the rock before it even hits it, and head in the other direction, straight through the clay.

Moisture, nutrients, obstacles, dangers: the root cap was sensing them all, sorting and steering accordingly. Darwin called it a "root-brain." Slice off the little cap, and the roots still grow, but blindly—they will continue in whatever direction they were angled when the root cap was removed. But then, the miracle: a removed cap will begin to regenerate in a few days, exactly as it was before. One of plants' greatest strengths is that they can regrow virtually any amputated part, but when a leaf grows back, it always grows back differently. The root cap is the only part to grow back exactly the same.

"We believe that there is no structure in plants more wonderful, as far as its functions are concerned, than the tip of the radicle," Charles and Francis wrote in the book's last paragraph with unabashed glee. No matter what they did to the root cap, it reacted in kind. "It is hardly an exaggeration to say that the tip of the radicle thus endowed, and having the power of directing the movements of the adjoining parts, acts like the brain of one of the lower animals; the brain being seated within the anterior end of the body; receiving impressions from the sense-organs, and directing the several movements."

We tend to think of science as a steady progression toward truth. Had this root-brain hypothesis been true, one might think, this radical new view of plants would have taken hold and immediately rerouted science down the path of viewing plants as animal-like in their capacity to direct their lives. But science's biggest flaw and biggest virtue is that it almost always mistakes agreement for truth. And no one agreed with Darwin. He was roundly rebuked by contemporary botanists. The root-brain hypothesis was promptly forgotten for the next 125 years, and to this day, we still don't know if it is true or false.

In *The Structure of Scientific Revolutions*, Thomas Kuhn draws the history of science not as a picture of linear progress, with new discoveries building on old ones, but rather as a series of abrupt paradigm shifts within specific fields, in which a set of conditions align to provoke a scientific crisis and a shift from one system of thinking to a completely new one. The crisis, here, is the important part. "Normal science" is the way of doing science that prevails before a crisis. Normal science is necessarily hostile to anything that falls substantially far outside it. One might think of how Copernicus and Galileo were received for their belief that Earth revolved around the sun, or Darwin for proposing evolution in an age of God's Will. Louis Pasteur was faced with extreme resistance from the medical community for supporting the germ theory of disease. The list of scientific luminaries who were punished before their theories were accepted is a long one. "No part of the aim of normal science is to call forth new sorts of phenomena; indeed those that will not fit the box are often not seen at all," Kuhn wrote.

A paradigm can't ask questions about something it doesn't see as existing in the first place. The resistance by scientists to scientific discovery is a known fact; it serves as a bulwark against quackery. But it also often misses or delays actual discoveries. The recognition of something as a significant anomaly that needs explaining—as Ian Hacking put it in his introduction to Kuhn's book—is a "complex historical event." And even that is not enough to prompt a scientific revolution. There must be another paradigm to accept before a rejection of the first can take place. "To reject one paradigm without simultaneously substituting another is to reject science itself," Kuhn writes.

Accepting the idea of plants as intelligent beings—even in some way conscious—would certainly amount to a paradigm shift. Yet getting it wrong risks rejecting science itself, jumping into nothingness. The evidence—and then widespread approval of it—must accumulate first. The current situation in botany is a case study for a scientific revolution that has not yet met its resolution. Its resolution is not even guaranteed.* The scientific community is in the midst of reorganization; the basic paradigm of botany is in a state of transition. We have the chance, here, to see how scientific knowledge is made.

What happens after a paradigm shift takes place? Kuhn says everyone goes back to normal. It quickly becomes hard to believe that any other idea ever held sway. What started as a few agitated stones has provoked an avalanche, and there is nothing to do but join the flow. In fact there is only the flow. Most everyone who was originally hesitant embraces the new paradigm as though it were always obvious, natural, preordained. I wonder if this will happen with plants. Will we, I wonder, look back in forty years and see our earlier beliefs about

---

* A scientific community in the midst of a crisis, Kuhn writes, will engage in "extraordinary" research, rather than "normal" research, and will see a "proliferation of competing articulations, the willingness to try anything, the expression of explicit discontent, the recourse to philosophy and the debate over the fundamentals." When I first read that line, it startled me; it fits the current state of botany so precisely.

plants as being as absurd and untrue as we now see they were about vivisection?

Eventually, Kuhn says, only a few elderly holdouts will remain. "And even they, we cannot say, are wrong," Kuhn writes. After all, they were right within the phase of scientific history they still cling to. But it's a new world now. "At most he may wish to say that the man who continues to resist after his whole profession has been converted has ipso facto ceased to be a scientist." They are uninvolved, out of step, left behind.

In 2006 a group of plant scientists tried deliberately to start a small but unignorable avalanche, in the hopes that it would change the paradigm. In a controversial article, they accused scientists of engaging in witting or unwitting "self-censorship," spooked into silence by the long specter of *The Secret Life of Plants*. The stigma had inhibited asking good questions about possible parallels between neurobiology and phytobiology, and "perpetuated ignorance" of great scholarship—namely Darwin's root-brain hypothesis, which they wished to revisit.* The new group, largely composed of scientists well into their careers, called for the pursuit of the idea of plants as intelligent beings, in the sense that they could process multiple forms of information to make well-informed decisions. Each of them had backgrounds in watching plants do this very thing, and they'd grown tired, it seemed, of the linguistic attempts to dance around what was actually happening: the plants were acting intelligently. They named themselves the Society for Plant Neurobiology. František Baluška, a cell biologist at the University of Bonn, Elizabeth Van Volkenburgh, a plant biologist at the University of Washington, Eric D. Brenner, a molecular biologist at the New York Botanical Garden, and Stefano Mancuso, a plant physiologist at the University of Florence, were among the founding members. Our understanding of plants is still so

---

* I think of Kuhn again: "Often a new paradigm emerges, at least in embryo, before a crisis has developed far or been explicitly recognized."

crude as to be rudimentary, they said. "New concepts are needed and new questions must be asked."

Invoking neuroscience was a bold move—and plenty of botanists I've spoken to, more than a decade later, still think it was far too bold. But they were trying to make a point. Of course plants don't have neurons or brains. But research was suggesting they might have analogous structures, or at least some physiology that could do similar things, and a cognitive capacity that deserved to be taken seriously. Plants produce electrical impulses, and seem to have nodes at the tips of their roots that serve as local command centers. Glutamate and glycine, two of the most common neurotransmitters in animal brains, are present in plants also, and seem to be crucial to how they pass information through their stems and leaves. They have been found to form, store, and access memories, sense incredibly subtle changes in their environment, and send highly sophisticated chemicals aloft on the air in response. They send signals to different body parts to coordinate defenses. Plant neurobiology "aims to study plants in their full sensory and communicative complexity," they wrote.

And what is a brain, really, other than a hunk of specialized, excitable cells, coursing with electrical impulses? "Plant neurobiology" was nonliteral, sure, but it wasn't a stretch, its proponents said. We don't need new words for things that are functionally similar—just new prefixes. Plant brains, plant synapses, plant thought. See, they said: Darwin was doing it a century ago.

Sometime after the age of the philosopher-naturalists, of Humboldt and Darwin, science became a pursuit of specialization. Despite relatively recent gestures toward interdisciplinary academics, we still live in an era of specialists, each of whom sees only their own narrow band within the larger problem of how life works. This has produced tremendous leaps in knowledge; with specialization comes depth. Yet for the most part, each specialist remains unaware of the larger picture. Maybe when dealing with plants, this is a formula for ignorance; a plant is a multidimensional organism in constant biological conversation with its surroundings, the bacteria, fungi, insects, minerals, and other plants

that make its world. It is no wonder then that zoologists and entomol-
ogists have been the ones to make some of the most groundbreaking
discoveries about plants, often by viewing them through the lens of
animals and insects. That's not to knock botanists, but in an age where
genetics dominate, many have ceased to see the plant as a pulsating
whole, and instead see it as an amalgam of genetic switches and protein gates. Of course, a human could also be seen in such terms. But
what is missed by looking that way?

The Society for Plant Neurobiology eventually backed away from
their provocative name; they became the Society of Plant Signaling
and Behavior. Yet even the word *behavior* still caused some botanists
to prickle. Pandora's box was already opened. What came next were the
rebuttals. Extremely salty ones.

Academics, armed with hyperliteracy, can be vicious when they dis-
agree. In the pages of *Trends in Plant Science* (TiPS), I read as skeptical
researchers slung thinly veiled academic venom. One researcher described the whole incident to me as the "TiPS Ruckus," and told me
about versions of letters from colleagues that were never published, or
where their antagonism was at least taken down a notch before publication. But one section of a letter written by the anti-intelligence camp
seemed to me to be especially telling. "Although Darwin got a lot of
things right, his brain analogy simply does not stand up to scrutiny,"
wrote Lincoln Taiz, the author of the textbook *Plant Physiology*, in a
letter coauthored by him and several colleagues. "If the root tip is a
brain-like command center, then so, too, is the shoot tip, the coleoptile
tip, the leaf, the stem, and the fruit. Because regulatory interactions are
occurring throughout the plant, we could regard the entire plant as a
brain-like command center, but then the brain metaphor would lose
whatever heuristic value it was originally supposed to have."

The comment was meant to be dismissive. Instead I thought it revealed a failure of imagination. Perhaps the entire plant *could* be regarded as a brain-like command center. What then? I thought of the
octopus, with its brain-like arms, the neurons distributed throughout
its body. We are just beginning to imagine what the world looks like

to them. There's no doubt it looks entirely different than it does to us. There is also no doubt that their distributed neuronic substrates are part of what gives them the capacity for such intelligent behavior, as well as the distinction of consciousness we have so recently deigned to bestow upon them. Seeing plants this way would add them to the conversations just bubbling up about different forms of distributed intelligence, too; the idea that decentralized networks built by fungi and slime molds can be intelligent, and perhaps even more agile in their ability to react to new challenges precisely because of their diffuse nature.

Even the human brain, as centralized a processing center for our bodies as it may be, appears rather less clearly centralized within itself. When neuroscientists peer inside the brain, they find a distributed network. No discernible command post exists. Our own intelligence appears to emerge from a network of specialized brain cells exchanging information, but they do not appear to answer to some governing force. The intelligent decisions we make emanate not from one specific place but from a sort of network, a consolidated city of interconnected, communicative parts in our skulls.* As

---

* The questions this poses for consciousness reverberate yet more loudly. Is there also then no ghost in the machine? The question of human consciousness is mostly an ongoing debate between two camps. The first are those who believe we owe our consciousness to a force beyond the material workings of our brain—something like a soul, let's say, or an as-yet-undiscovered property beyond or apart from our physical brain matter. Panpsychists fall under this camp. The second broad camp are those who think consciousness is purely a biological phenomena borne of evolution just like everything else in nature, and that its cause is most likely located in the immeasurable complexity of the organ in our skulls; we just haven't discovered the mechanism yet. This camp are sometimes called the materialists. But neither broad theory technically requires a brain, at least not in exactly the way ours happens to exist. In fact, both could leave room for the possibility of consciousness in degrees, rather than something a creature has or does not have. If consciousness is the result of a transcendent, free-floating property of the universe, could a creature not have more or less of it than its neighbor? And if consciousness is simply an emergent property of biological evolution, couldn't that trait simply have been more or less emphasized in the evolutionary trajectory of each creature? When it comes to

journalist Michael Pollan once put it, there may be no wizard behind the curtain.

New ideas in science provoke new methods, and new theories. Without revolutions, science would degenerate. It's important to keep this in perspective. Kuhn said that a scientific paradigm shift has the power to change one's view of the world in which we live. "Of course, the world itself stays the same," Kuhn wrote. Plants will go on being plants, whatever we decide to think of them. But how we decide to think of them could change everything for us.

---

plants, the question at hand is whether consciousness could appear in something other than ourselves and a select few animals at all.

Chapter 3

The
Communicating
Plant

I awoke in the mornings at dawn because I'd recently noticed that this was when the world was most alive. How had I ever not known? At dawn everything shimmered with activity. Full daytime was a dead period in comparison. The birds in the salt marsh below the house called wildly, like they were caffeinated. I was not, at least not yet, but I liked the bleary minutes before my mind turned to more human affairs. I was at a writing residency on Point Reyes, in California, to continue thinking and writing about plants. I wanted to get clear on what questions to ask, how to organize my curiosity. The little group of us lived right on the edge of the San Andreas Fault, perched on the edge of one great tectonic plate looking across the marsh to the other. The yard was planted in salvias, representatives of the sage family, in my mind the reigning monarchs of scented plants, all wafting camphor and spiced oil. There were great big specimens of desert purple sage, silvery blue sage, and a Copper Canyon daisy with its thousand gold blooms. Thick bushes of glossy rosemary were pricked everywhere with ice blue flowers that closed their winged petals at night.

I stepped out onto the porch. The beard lichen that clung to a white birch beside me looked as though it had slinked with great purpose up the young tree, sheathing its trunk like a tube sock. It was advancing scruffily too along the lower branches. I had the dreamlike sense that it had cunningly frozen in place just in time for my glance. Lichen time is slower than human time, so I supposed it had—it and all its brethren, in motion but frozen in our moment of notice. I ankled out like a deer, smelling the air, picking across the scrubby lawn as though any sudden movement could unsettle the veil of scent. It didn't—of course it didn't—these smells were not for me. I'd read about new theories of plant language the night before that were no doubt working on my predawn mind. The language of scent, they said, was wafting messages on the air. I began to understand that a many-layered drama was playing out all around me, with more characters and plot lines than a Russian epic. Some of these I could smell, and there were many more my nose was too naive to notice.

I decided to start here: Were plants communicating? And what would it change if they were? Communication implies a recognition of self and what lies beyond it—the existence of other selves. Communication is the forming of threads between individuals. It's a way to make one life useful to other lives, to make oneself important to other selves. It turns individuals into a community. If it is true that a whole forest or field is in communication, it changes the nature of that forest or field. It changes the notion of what a plant *is*. What is a plant without a means to communicate? A husk. And without conversation, a forest is not a forest.

I had spent the night before reading the paper that changed botany forever. It had by now been mostly forgotten, and seemingly missing from the digital record; Richard ("Rick") Karban, a plant and insect ecologist at the University of California-Davis, had to mail a photocopy of it to my house. I'd learned that it led to fights that ended at least one scientist's career, and cracked open the question of plant communication for all future botanists. But for all the changes it would usher, the paper I was reading was obsequiously mild. Its language was reserved to the utmost. This made sense. It existed on a knife's edge; its conclusion was so novel, it would be easier to dismiss than embrace. And dismissing was the norm in the time that it was written. Things in plant biology were tense. David Rhoades would have to tread lightly.

It was 1983, and the repercussions of *The Secret Life of Plants* were still being felt. David Rhoades—Davey to most—was a zoologist and chemist at the University of Washington who mostly studied insects. He was British, gregarious, portly, and given to chain smoking and gesticulation. His thick mustache crept past the corners of his mouth, and his eyes closed when he laughed. He took his data deadly seriously, and loved designing experiments with the absolute least expense, rigging lures for insects out of grocery store finds. His paper would change everything, and in a cruel twist, it would end his career. Because back then, no one believed him.

The paper, published in *Plant Resistance to Insects*, a relatively obscure (if you can believe it) publication by the American Chemical Society, wrapped its incendiary offering in felted layers of science-speak. For twelve full pages Rhoades dutifully copied out pupal weights of caterpillars and leaf losses from trees. He'd been watching the university experimental forest get decimated by an invasion of tent-forming caterpillars for several years, he explained. But suddenly something had changed; the caterpillars began to die. Why, he asked, did the voracious caterpillars suddenly stop munching, leaving tree leaves intact? Why did they seem to abruptly die out?

The answer, Rhoades discovered, was improbable, remarkable, and dangerous: the trees were communicating with each other. Trees the caterpillars hadn't yet reached were ready; they'd turned their leaves into weapons. The caterpillars that ate them got sick and died.

Communication between trees via their roots had been established somewhat earlier than Rhoades's finding, but this was different. The trees were too far apart to be passing information through their roots. The message—that the caterpillars were coming—was getting through regardless. His excitement at what that implied was more than he could dampen. When all room for dry description was exploited, and there was nothing left to do but say it, Rhoades could not help but let the true nut of the paper out as a chirp, using that most exposing punctuation: "This suggests that the results may be due to airborne pheromonal substances!" The trees were signaling to each other, he said, across long distances, through the air.

Communication is another of many basic life processes with no agreed-upon scientific definition. For most of us communication is about something we express in order to tell another being something they need to know. It suggests a complicated form of intentionality, forethought, and an awareness of cause and effect. That may be so, but life began to communicate, depending how you define it, prior to the advent of more complex existence. It began with the first multicellular organism at least six hundred million years ago. For life to open to the possibility of multicellularity, individual cells had to coordinate among themselves. Up to that point, all life was single-celled. These autonomous little selves were adrift in the ancient sea, each making their way on their own. For more complex forms to emerge, individual cells had to share information with each other.

To this day, in order to coalesce into a body, each cell in an organism must know who it is and what it does. Cells understand themselves by way of other cells; in a chain of three cells, for example, the third cell knows it is third—and thus endowed with a special task reserved for third cells—because of its awareness of the presence of cell one and cell two. That is the nature of a self-organizing system, a cohesive body. But *how* that cell knows it is third remains a mystery. We know that information must be passed to it by its collaborator cells. This communication, whatever it is, begins with the very first cell division, when one cell becomes two and then four, which is the strategy every multi-celled organism uses to grow. The medium of that information—electrical? Chemical? Some other form?—is unknown. The nature of the communication also remains a major question in animal embryology; we'd like to know how a sperm and an egg self-organize to make us.

Plant cells do this too. In the most liberal use of the term, they "talk" to each other. In this way, each cell understands what it is for—or, put another way, who they are. Barbara McClintock, the Nobel-winning geneticist who discovered genes could shift their positions in corn, called this cellular awareness the "knowledge the cell has of itself."

When cells talk, big things happen. All plant life proceeds from these foundational interactions. In 2017, researchers at the University of Birmingham identified the presence of a "decision-making center" within dormant seeds that integrated information and decided when the plant should emerge. This center is made up of a cluster of cells located at the tip of the seed's embryonic root. The cells communicate to each other about the abundance of two hormones in the seed; one hormone that promotes dormancy, and one that promotes germination. The cells integrate information about the changing temperature of the soil around them to regulate each hormone. In this way, the cluster of cells decides when to flip the switch and emerge into the world. The timing of the decision to emerge is critical. This risky decision is made more accurate by relying on the cumulative responses of multiple cells; by making a decision based on two opposing variables—the relative abundance of two hormones, both of which are sensitive to temperature changes—the plant has a higher chance of making a good choice in a fluctuating world. This is, the researchers noted, a method of cell-to-cell communication analogous to certain structures in the human brain. Our brains also pass antagonistic hormones back and forth between cells to improve our decision-making in a fluctuating world. Rather than making the decision to move a muscle based on a single input, the brain makes its decisions by accumulating hormonal information from separate cells, and weeding out irrelevant information in the process. This is at its heart a case of cells in communication.

Through the chatter of their cells, plants are self-organizing systems. But that whole plants might be considered to communicate intentionally with each other—that communication could extend beyond one plant to others—is a relatively new and still controversial concept in botany. One key problem inclines the whole thing toward debate: there is no agreed-upon definition for what counts as communication, not even in animals. Does the signal need to be sent purposefully? Does it need to provoke a response in the receiver? Much as *consciousness* and *intelligence* have no settled definition, *communication* slip-slides between the realms of philosophy and science, finding secure footing in neither. To proceed with a sense of clarity, I'll define communication as happening when a signal is sent, is received, and causes a response. You'll notice I didn't say when a signal is *intentionally* sent. Intentionality is more difficult to discern, in part because we don't know what it's like to be a plant. Intention poses the hardest of problems, because it cannot be directly discovered. We can only build knowledge around the problem of intention, drawing our perimeter closer to it, and hope that in encircling it, its form might begin to take a shape we can understand.

Yet at first, just the mere idea that plants could have any information to convey to each other was absent from the scientific map. All Rhoades knew was that a plague began, and then stopped. In the spring of 1977 the experimental forest at the University of Washington was in its third consecutive spring of a sustained and gruesome tent caterpillar attack. Red alders and Sitka willows, usually able to withstand the nuisance of a few months' tenancy by these web-weaving leaf-eaters, were succumbing by the hundreds. The caterpillars were almost completely defoliating them, which is to say they were visiting a famine upon them; a tree without leaves to photosynthesize in the growing season cannot make sugar and effectively starves to death.

But the following spring, in 1978, the balance of power seemed to shift. The tent caterpillars were dying now. Their populations crashed. Few caterpillar eggs appeared on the remaining tree leaves, where the previous spring they were everywhere. Eggs that did show up didn't tend to hatch. By the spring of 1979, the caterpillars had vanished entirely. The trees stopped dying. Their leaves were full and productive. The fortunes of each party had reversed.

As every ecologist knows, nothing changes in an ecosystem without a reason. Something drove that shift. Rhoades, who had a PhD in both organic chemistry and zoology, began looking for an explanation. For years he'd been cultivating a provocative idea, without much support from his colleagues: he thought plants might develop resistance to certain threats after being exposed to them, much as the animal immune system builds up antibodies to a disease the animal has already encountered. He'd noticed that insects would often start to eat a plant and then eventually stop, despite there being plenty of good leaves left to eat. Again, nothing occurs in nature without cause. Something made the insects stop.

Might the plant be clocking the invasion, and mounting a sort of immune response to it? That would explain the lag; plants operate on a slower timescale than insects, so it made sense that they'd react more slowly too. His lab tests bore this out. He saw that after leaves had been enduring a swarm of munching caterpillars for a while, their chemistry changed; the plant would alter the contents of its leaves to be less nutritious. The idea that a plant would actively defend itself, though, was heretical to the whole premise of how scientists thought plants worked. Plants were not supposed to be that active, or have such dramatic and strategic reactions. Rhoades found few supporters for his hypothesis.

But the tent caterpillar invasion on university grounds provided the perfect scenario to study his theory in the real world. The besieged trees did eventually change the composition of their leaves, sickening the caterpillars, who would essentially die of starvation through diarrhea. He was pleased. His theory checked out. But he also noticed something else: even the leaves of faraway trees, which the caterpillars had not yet touched, changed their composition too. They'd been warned, and somehow the warning had traveled a long distance. Plants are tremendous at chemical synthesis, he knew. And certain plant chemicals drift on the air. Everyone already knew that ripening fruit produced airborne ethylene, for example, which prompts nearby fruit to ripen too. The commercial fruit industry used it to ripen warehouses full of unripe bananas just in time for sale, making the global trade of an otherwise fast-rotting fruit possible. It wasn't unreasonable to imagine that plant chemicals containing other information—say, that the forest was under attack—might drift through the air too.

Rhoades presented his hypothesis at conferences. The story of the talking trees spread, whispered from botanist to botanist like arboreal gossip. Could it be true? But none of his peers were willing to take the risk of publishing something so outlandish. The discovery ended up buried in an obscure volume. Rhoades spent the next few years performing the usual academic duties. He taught and guest-lectured, all while being bludgeoned by colleagues in journals and at conferences. He turned more and more toward his role as a mentor, finding considerably more openness in students and new professors, perhaps because they had not yet been blindered by institutional conservatism.

Rhoades began to exchange letters with Rick Karban, a newly minted entomology professor who was interested in his idea of "induced resistance," the phenomena in which a plant chewed by insects would alter its chemistry and be less likely to be munched on later. Karban thought the usual view of plants at the whims of their environment couldn't be right. He'd come up studying cicadas, which lay their eggs in trees. When the larvae hatch, they drop to the ground, burrow into the tree's roots, and stay there for seventeen years, sucking its sap. To the tree, it's a huge nuisance to have all that nutrition leak out of its lower parts before reaching its upper ones. As a young scientist, Karban read a paper by pioneering cicada researcher JoAnn White, who discovered that some trees were able to locate the place on their branch where the cicada eggs sat and grow a callus around them, suffocating them to death before they could hatch.

Karban, like Rhoades, thought that plants couldn't possibly be passive. Karban invited Rhoades to speak to his graduate class. After that, they stayed in touch; Rhoades read Karban's manuscripts and commented on his grant proposals. But Rhoades's own life was crumbling around him. The reprimands continued. He had trouble replicating his own study. He tried for two years; sometimes it worked, sometimes it didn't. After a pattern of rejection, Rhoades gave up applying to grants, the researcher's equivalent to giving up eating. Finally he left the world of scientific discovery. He took a job teaching organic chemistry at a community college, and opened up a motel on the Pacific coast. He was diagnosed with terminal cancer in the 1990s and died in 2002. With his work, he'd arrived at the right place, only it was the wrong time.

But all the while, the tide, at least for others, was slowly turning. Six months after Rhoades published his paper, Ian Baldwin and Jack Schultz, then young researchers at Dartmouth College, published a very similar finding. It isn't always clear why fortune favors some and not others in the arc of scientific history. In this case, it's likely a combination of luck and study design. Their work was done in the safety of a lab. The outdoors are a messy place to do science; lab work is clean, controlled, specific. Baldwin and Schultz placed pairs of sugar maple seedlings inside the sterility of a growth chamber. The seedlings shared the same air but didn't touch. Then the researchers ripped the leaves of one and measured the response in the other. Within thirty-six hours, the untouched maple seedling loaded up its leaves with tannin. In other words, despite not experiencing damage itself, the untouched maple went to work making itself extremely unpalatable.

Baldwin and Schultz noted that they were the second in line to notice this phenomena, acknowledging Rhoades in their paper. They even went so far as to use the word *communication* in their work (Rhoades never used the c-word, choosing to dance around it). The mainstream press understandably seized on the wording, printing headlines about "talking trees" in national newspapers. They were generally chastised by peers for using such human language for plants, but in contrast to Rhoades, to say their careers recovered would be an understatement. Today Baldwin is one of the most successful and prolific individuals working on plant behavior. He has a large team of graduate students and postdocs working out how tobacco plants, in effect, communicate, defend themselves, and choose which other tobacco plants to have sex with. Jack Schultz spent decades as a major contributor to the field of communication between plants and insects, and was known to say that the scent of cut grass is the chemical equivalent of a plant's scream. Both say they were inspired by Rhoades.

Years after Rhoades's death, Jack Schultz said he thought he knew why Rhoades could never quite get the trees to do it again; it is now known that along with a host of other dramatic seasonal changes that trees go through, the airborne chemicals they produce are seasonal too.

The original study was done in the spring, and Rhoades was trying to replicate the study in the fall. No wonder the result changed. The trees were in a different phase of their yearly cycle. He wasn't on the wrong track; there were just more variables yet hidden from his view.

Rhoades reminded me of Gregor Mendel, the Augustinian friar and father of genetics, who tried to replicate his beautiful pea-crossing studies in hawkweeds. It never seemed to work; he died frustrated and defeated, believing his life's work to be irreproducible and therefore meaningless. Of course it was anything but that. What he didn't know was that hawkweeds have a strange proclivity: they can produce seeds at random without pollination. In other words, they periodically clone themselves instead of reproducing through plant sex, confounding the whole process of studying genetic crossing. Nature, never a flat plane, has always more folds and faces still hidden from human view. The world is a prism, not a window. Wherever we look, we find new refractions.

AROUND THE SAME time as Rhoades and Baldwin and Schultz were defending their papers, outside the halls of botany a South African wildlife manager was making what can only be called an anecdotal assessment. It isn't a peer-reviewed experiment, but I heard it repeated enough times—including by the South African wildlife manager himself—to feel it was worth enumerating here, with all the proper caveats. I take it for what it is: a story.

In 1985, Wouter van Hoven was in his office in the zoology department at Pretoria University when he got an unusual call from a wildlife warden. In the last month, more than a thousand kudu, a particularly majestic species of antelope with elegant stripes and long, curling horns, had dropped dead on multiple game ranches in the nearby Transvaal region. The same thing had happened the winter before. In total some three thousand kudu had died. Nothing seemed wrong with them, no open wounds, no disease, though some looked a little thin. Could he come out as soon as possible? The ranch owners were beside themselves. Van Hoven was a wildlife nutrition zoologist who specialized in African ungulates. He should be able to figure this out, he thought. He'd be over right away.

When Van Hoven got to the first game ranch, dead kudu were lying about as if a war had just been fought. But the first thing he noticed after the stench was that there were too many of them for a ranch that size. As a rule, there should not be more than three kudu per 100 hectares, and this ranch had about fifteen per 100. The same was true at the next few ranches he visited. Game-ranch hunting had exploded in popularity, and to cash in, ranchers were pushing the limits of their land.

He opened up several kudu and saw stomachs full of crushed acacia leaves, undigested. He looked out at the giraffes, who were spread out along a swath of savanna, nibbling acacia trees and evidently not dying.

After a few weeks a picture began to come together: when acacias begin to be eaten, they increase the bitter tannin in their leaves. Van Hoven already knew this. It's a gentle defensive mechanism. At first, the tannin rises just a little. It's not dangerous, but it tastes bad. Typically, that's enough to deter a kudu. But both of the last two winters were extremely dry. All the grass was dead. Too many kudu, penned in by game fences, had nothing else to eat and nowhere else to go. He figured they had continued eating the acacia leaves, despite the bitter taste, because they had to. He pulled out a few clumps of chewed acacia leaves from a kudu gut and brought them to a lab.

Kudu, Van Hoven knew, could handle about 4 percent tannin content in a leaf. Above that is trouble. The acacia, he figured, kept raising the level of tannin in the leaves, tit for tat. The kudu kept eating. And then, clearly, the acacias delivered a lethal dose. The undigested leaves Van Hoven tested from the kudu's stomachs were 12 percent tannin.

The way he sees it, nature had basically decided, "I'm going to have to reduce the population of these animals," he says. "And then it did it."

Van Hoven remembered reading about chemical signaling between trees a few years earlier, likely Rhoades's or Baldwin and Schultz's paper. With that in mind Van Hoven broke some acacia branches and sampled the air. Sure enough, the damaged trees were releasing great plumes of ethylene, certainly enough to waft over to a nearby tree. Surrounding trees were being alerted and changed their behavior accordingly, he decided. It was a coordinated poisoning.

He went back to the giraffe. How did they get away with eating the acacia leaves? "They'll eat and eat, and all of a sudden they'll stop and walk away. Even though there's plenty of leaves." It didn't make sense from an energy-conservation perspective. But soon it was clear they were only eating from one out of every ten trees, and never downwind. He guessed the giraffes had learned to eat only from trees that had not received a warning to release their tannin.

RICK KARBAN DOESN'T like that story, or the way it gets repeated. He's spent his career tussling to get unconventional work published, but hasn't lost faith in the arduous peer-review process as an essential safeguard against false paths. Without it, science loses all credibility. Everyone needs the counsel of their peers to keep the threat of human error in check. And the kudu story hasn't undergone that kind of review. Other botanists across the spectrum of belief respect Karban mightily, even when they themselves are not the type to entertain ideas of plants doing much of anything with intention. But when they talk about Karban, they use words like "rigorous," and tell me I should go watch him work.

Karban is a lithe beanpole of a man, with arrow-straight posture and feathery white hair. On the day I find him in his office on the third floor of the University of California–Davis biology building, he is wearing bright-orange tennis shoes and using a yoga ball as a desk chair. It's 12:00 p.m., and the bird clock on the wall behind him squawks. "It's old, the birds aren't right," he says by way of explanation for why a bluejay squall just came out of a finch.

Karban's office is a small rectangle set off from a large, open-plan entomology lab where Tupperware containers of tiny dead butterflies litter a bench, and two insect nets on long poles, each bigger than me, lean against a wall. I ask him what a botanist is doing in a bug lab. He shrugs. "I started out in cicadas," he reminds me, and most of his work still exists at the place where plants and insects meet. His field site for the last twenty years has been nestled on a mountainside in Mammoth Lakes, California, a gorgeous moonscape of subalpine forest and sagebrush desert high in the mountains. We set off for there.

The Valentine ecological study area in Mammoth Lakes, owned by University of California–Santa Barbara, is a 156-acre reserve situated in the caldera of an ancient volcano, eight thousand feet above sea level. There's no fence to keep tourists out, just a sign warning that trespassing won't be tolerated. Most wouldn't even know where to look; the entrance is a perimeter of scraggly pine forest with no trail through it, unappealing compared to the ski area practically next door.

But immediately beyond the hedge of trees, the land opens into a rise that, in July when I go to visit, is covered with frosty green sagebrush and glossy manzanita crowns. Giant Jeffrey pines, armored in scales of vanilla-scented rust-orange bark, stand above the low plants. Corn lily, pale pink phlox, white rein orchid, mule's ears, serviceberry, and orange tufts of semiparasitic desert paintbrush emerge from the bone-dry gravelly ground. Two deer, young bucks with nubs for antlers, leap away as I walk. So do grasshoppers. Above the ground-level spectacle rise the jagged tops of Sierra Nevada peaks, still smudged with snow despite the July sun.

And then there's Karban, bent over a sagebrush bush, plucking off tiny black beetles with tweezers. He hands me a pair of tweezers and a paper pint container—the kind used for ice cream, this time with air holes—and tells me to start collecting the bugs, which he'll reuse in future experiments. As a science journalist I never tire of discovering anew how arts-and-craftsy field research can sometimes seem. He'd placed the beetles on the bush himself the night before; whether they were still there would tell him how hard the plant had tried to get rid of the predator.

But beetles have predators, too.

"Ah, a ladybug is eating one," Karban says, momentarily disappointed at the lost data point. "Ah, okay! It's real life!"

Karban's research has shown how chemicals wafting off sagebrush can be interpreted even by nearby wild tobacco, and how that same wild tobacco, when it begins to be damaged, can summon predators to eat the caterpillars that feed on it. He's also found that sagebrush are more responsive to cues from their genetic kin. If a sagebrush receives a chemical signal through the air, perhaps one that indicates dangerous predators are nearby, it'll be more likely to heed the warning if it's from a close family member.

At the time of my visit, Finnish evolutionary ecologist Aino Kalske, Japanese chemical ecologist Kaori Shiojiri, and Cornell chemical ecologist André Kessler had recently found that goldenrods that live in peaceful areas without much threat from predators will issue chemical alarm calls that are incredibly specific–decipherable only to their close kin–on the rare occasion they are attacked. But goldenrods in more hostile territory signal to their neighbors using chemical phrases easily understood by all the goldenrod in the area, not just their biological kin. Instead of using coded whisper networks, these goldenrod broadcast the threat over loudspeaker, so to speak. It is the first time research has confirmed that these sort of chemical communications are beneficial not only to the plant receiving them but also to the sender.* When times are truly tough, you don't want to be left standing in a field alone when it's over, if you're a plant. There'll be no one to mate with, no one to help bring in pollinators. It's the closest scientists have come to showing intentionality in plant communication: these are signals meant to be heard. And as we know, by some measures, intention is an indicator of intelligent behavior.

Over and over, Karban has found ways to use methods from the

---

* Prior to this goldenrod finding, one interpretation of plant communication was that it wasn't communication at all. Rather, plants had just learned to eavesdrop on the signals their neighboring plant was sending to its own disparate branches, to tell other parts of its body to mount a defense. As one researcher put it, volatile signaling wasn't communication, it was a "soliloquy." But this goldenrod paper upset that notion. Everyone in the conversation seemed to benefit from the exchange. See Martin Heil and Rosa M. Adame-Alvarez, "Short Signalling Distances Make Plant Communication a Soliloquy." *Biology Letters* 6, no. 6 (2010): 843–45.

world of animal behavior research on plants—and over and over, they seem to work. Karban remembers a finding about songbirds that he thought might apply to what was happening with the sagebrush. He tried to replicate the Finnish paper to find out. It worked. Sagebrush also use "private" means of communication to warn only their family groups about insect attacks when the threat from bugs was generally low. Basically, they are using backchannels—the chemical compounds they use are complex and specific to them and their closest allies. But when the whole community is being heavily attacked, sagebrush will switch to "public" channels, emitting more universally understandable alarm calls. This tracks perfectly with something that has been known about songbirds for a long time. In peaceful places, where relatively few dangerous predators lurk, birds use extremely specific song phrases to warn only their family group that something is wrong. But when the birds are facing widespread danger, they switch calls, making alarm sounds that everyone in the area can understand, even members of other bird species. This makes sense, again, in terms of community survival; when the whole neighborhood is threatened, it's best to save as many of your kind as possible, regardless of if they're family or not.

I thought about what this meant for plants. Now that this has been found in more than one species, it is safe to assume it will be found in others, and may even extend throughout the plant kingdom. That means plants could be said to have dialects, and are alert to their contexts enough to know when to deploy them. More than that, they have a clear sense of who is who; who is family, and who is not. They are in touch with their surroundings, and with the fluctuating status of their enemies. Their communication is not just rudimentary but complex and layered, alive with multiple meanings.

The capacity for variation brings plants closer to us in crucial though still crude ways. Different changes in our lives prompt different responses in us, of course. We assess threats and tailor our reactions to fit them. But this made me wonder about variation among individuals: humans are not all the same, and our threat responses are so personal to each of us. Levels of bravery or fear, daring or caution, vary wildly between us all. I didn't suppose human concepts like "fear" would apply directly to plants, but a more conservative question still seemed worth asking: Do individual plants have this sort of spectrum of reactions too?

I was delighted to learn that Karban's latest experiments concerned exactly that: he wants to know if plants have personalities. Personality research entered the world of zoology relatively recently; in the last 20 years, animal science has begun to take seriously the idea that individual animals have personalities—consistent, unique ways of responding to the world—and that they are worthy of study.

Karban often talks to colleagues who study personality traits among animals, and he's come to a simple yet groundbreaking conclusion about how to approach his research: animals and plants are clearly different, yet they share a common world. Their daily travails are very similar. They need to find food, and they need to find mates. And they need to do it all while other things are trying to eat them. "If animals have solved a problem in a particular way, it's not unreasonable I think to ask, huh, I wonder if plants have done something analogous."

Usually, when scientists measure characteristics of organisms—whether plants or animals—they look at the average of the tendencies of the entire group. For at least the last hundred years in plant biology, individual plants within a species have been seen as replicants. No individual trait matters to science, which only looks at the average of the traits of the entire population. If one individual falls too far outside the average, it tends to get discarded from the study as an outlier. "What individuals do is seen as just noise," Karban explains. But his work with sagebrush throws away the relevance of averages. Personality research treats individual differences as valuable data. Each one is a point on the spectrum of behavior. The noise becomes the signal. "This is the opposite approach: It's paying attention to the variation among individuals."

After a long career studying how sagebrush send signals to each other, Karban is finely attuned to variation in this process. He sees that the exchange doesn't turn out the same way every time. Sometimes a plant will signal distress, and its neighbors won't produce defensive compounds, or sometimes they just produce less. Karban believes this may be because individual plants may have different tolerances for risk—one metric of personality. Some, he says, might exhibit a personality akin to, say, natural-born scaredy-cats; they will signal wildly at the slightest disturbance. In that case, other plants in the same family will treat their scared kin like the boy who cried wolf and ignore them. They won't produce compounds of their own.

As we walk through the sagebrush meadow, we talk about our lives. Karban, I learn, is a New Yorker far from home. It turns out he grew up in the same Lower East Side apartment complex my mother lives in now. The Lower East Side in the 1960s was a rough place to be a kid, and for a boy that meant being willing to get into fistfights to defend yourself, or else lose your lunch money. That wasn't Karban's mode. "Risk-averse," he called himself. He didn't fit in, or at least always felt a certain separation from the world, a skepticism about its intentions toward him. A black sheep. So he spent a lot of time inside, wishing he were somewhere else, a gentler place, outside the relentless humanity of the city. As soon as he could, he moved to the opposite side of the country, to study the complexity of nonhuman creatures. He still insists on doing almost all of his research outside, in the messy reality of unpredictable ecosystems.

To be sure, Karban's personality work is the outer edge of plant behavior research. He can afford to do it; as a respected scientist with forty years of scientific research under his belt, his dedicated interest in the potential personalities of plants is a signal to anyone paying attention that this is a thought experiment whose time has come. If his results are compelling, and repeatable, they could have enormous implications—well beyond the tiny world of plant researchers. A diversity of human responses to the environment, one might argue, makes us more resilient as a whole. The same may be true of plants.

In 2017 Charline Couchoux, a behavioral ecologist at the Université du Québec in Montreal, emailed Karban with a proposition. She had an assignment for her PhD requirements to collaborate with someone outside her field, and she had a methodology to identify individual behavioral differences in animals that he could repurpose for plants. Couchoux had spent thousands of summer hours watching chipmunks in the woods at the Vermont-Québec border. She'd tagged each of the dozens of chipmunks with a different-colored earring, but by the time she was done watching them, she knew each of them on sight and behavior alone.

Chipmunks have distinct distress calls; one for when they detect an aerial predator, like a hawk, and another for a land predator. Some chipmunks, she said, would squeal all the time. "Some guys would be eating seeds, and a leaf falls on the ground. They panic and they make a call," she said, screaming their little heads off about an imagined bird of prey. Those were the shy ones. "Some guys just keep foraging." When she controlled for sex, social status, and age, there were still distinct differences in chipmunks' personalities that remained stable over time. Some chipmunks were risk-takers, others were not.

Of course, these distress squalls are heard by other chipmunks. What the other chipmunks choose to do with that information seems to depend on how reliable the squaller was. "The main idea is that if you have some guys who cry wolf all the time, they shouldn't be trusted." She recorded calls from a range of chipmunks who fell on different places along what she and her colleagues called the "shyness-boldness continuum," and played them back to other chipmunks. The listeners perked up and paid attention when they heard a distress signal coming from a bold chipmunk, and seemed not to care as much about one from the oft-panicked.

From an evolutionary, survival-of-the-fittest perspective, one might think the shyer chipmunks were doomed. But Couchoux found that wasn't the case. The less aggressive individuals took less risks, so they ate less, and they had fewer babies each year. But they tended to live longer. Less risk-taking meant fewer chances to be gobbled by an eagle. At the opposite end of the spectrum were the really bold chipmunks. "They reproduce earlier, they eat a lot, they take more risks. They will have more babies, let's say three in one year. But then they die because they are eaten by a predator.

"Their strategies are different, but they can both work for life," Couchoux says. "This has been found in many species now, from big-horned sheep to fish." When it comes to personalities, there's a place for everyone.

At another field station 185 miles north of Mammoth Lakes, Karban keeps a field of ninety-nine sagebrush, and he knows each one personally. He and his graduate students have profiled each one genetically, and they know the relatedness of all the individuals. Already, they've proven that sagebrush are more responsive to cues from their genetic kin. Now, they've adapted their technique for sampling chemicals to study plant personalities.

To do this, he and his grad students damage a sagebrush plant, typically by snipping a few leaves. Then they put a plastic bag around the plant, to trap the volatile chemicals it inevitably pumps out. They use an oversize syringe to capture some of the chemical-laden air from within the bag. Then they spritz the air near another plant, and record its response. The next step is to formalize a personality profile for each one. Once they have that, they will be able to track the plant's responses over time, and see if their personality holds true over the course of their life. If a skittish bush stays a skittish bush, as Karban expects it might, then the field of plant personality research will truly be born, along with a clear way to study it.

As the study of plant communication evolves, new information is coming to light seemingly anywhere a researcher decides to look. Colleen Nell, a scientist at the University of California–Irvine, recently discovered that among the blooming desert shrubs commonly called "mule fat," female plants will listen to the cues sent by both male and female plants, but male plants will only listen to male plants. In other cases, plants appear to prefer to get their intel from family; one of Karban's studies found that sagebrush plants will listen to their genetic kin, but not genetic strangers.

This new research raises existential questions about what we think of as a healthy plant community, and what it means to actually protect them. Given these findings, simply growing plants is not enough;

if communication is a vital function of plants, then our care for them must also extend to protecting their ability to "talk" to one another.

Later, back at home and thinking about Karban's research, I would spend an eerie moment contemplating my house plants; were they being silenced? Are these companions who make my apartment feel more alive being deprived of some essential plantness? It seemed likely now. First of all, they were in pots. When it came to root-to-root communication, they were undeniably cut off from any connection to their fellow plant life—not to mention the network of fungi and bacteria they might normally associate with. But were they also cut off from the chemical form of plant speech? Did they breathe meaning into the air through chemical compounds, the way their wild counterparts almost certainly do? Almost all of the plants in my apartment were tropical varieties widely cultivated in nurseries. So far from their wild ancestors. Were these plants some diminished, domesticated version of their wild relatives, so many generations removed from the jungles of their kind that they forgot how to speak, perhaps never hearing their language? And lineage aside, was I now keeping them like animals in cages, confined, muted, in their pots? It was chilling to imagine. Or were they more like dogs to wolves, in need of my care, now that they'd lost the context for and traits of total self-sufficiency? I didn't know how to feel about that either. It was also, I knew, a flight of fancy. I'd probably let my mind travel too far. It's very easy to do that when thinking about plant agency. Yet—I scolded myself now, confusing things further—what is "too far" when the topic at hand is the agency of living creatures?

BACK AT MAMMOTH Lakes, now fully lying in the dry gravel dust in nylon khakis to get a bug's-eye view, Karban is counting beetles. His floppy field hat pokes up above the sagebrush bush he's buried his face in.

As I sit on the ground nearby, I inhale wafts of sagebrush's signature camphorous smell, herbal and slightly spicy. This is a bouquet of a few of the plant's many volatile chemicals, what they use to communicate with different parts of their body, signals that fellow sagebrush can eavesdrop on and respond to. That, Karban thinks, may be their version of "expressive" or "quiet," if we only learn how to listen.

Much like in humans, where the mind is studied by inference—what a person does—rather than neurological mechanisms, Karban is looking for patterns in behavior. "I'm a big fan of using what decades of psychology has learned, their methods, and asking if they apply to plants," he says. "In some cases that's no, and that's fine."

But he's found one method from the field of psychology research that truly seems to fit. The method helps researchers analyze behavior by separating it into two processes. The first is judgment, or the perception of raw information; the second is decision-making, or how one weighs the costs and benefits of different actions and selects the best one. This applies perfectly well to plants, he says. How different plants might weigh the threat of predators and then take action against them—by, say, making their leaves bitter, or in the case of tobacco, chemically summoning predators that will eat whatever is eating them—could be a strong signal of individual personality. How severely they judge the threat, and how they choose to act in response, could teach us volumes about the range of plant approaches to life.

As we walk out of the field site, we descend from the dry plain and into a shaded ravine with a stream running through it. Everything is intensely green. Karban points out a wild tiger lily, a cow parsnip. He spots a tuft of yellow-beaked monkey flowers. "When they think they've been pollinated, the stigma closes up. If it's really pollen, it stays closed. Like, okay, I've gotten what I'm after. But you can trick them with a blade of grass." He shows me, poking at the yellow bloom. "It will close, but then in half an hour or so it will be like, oh, that's not right, and open back up."

We keep walking. Quaking aspen, forget-me-nots, alders.

I ask Karban how all his work has changed how he sees plants. "People ask me, do plants feel pain?" he says in response. But the question misses the point. "Plants know they're being eaten. They probably experience it very differently than we do. They're very aware of their environment, they're very sensitive organisms. And the things they care about are very different from what we care about. They know when I am bending over them and casting a shadow. And it's ridiculous to think they'd prefer classical to rock," he says. "But they are sensitive to acoustics."

He becomes thoughtful, and pauses walking. "I have a lot of respect for them as—I don't know if conscious is the right word, but as very aware beings," he says. "That's new for me, in the last ten years or so. The facts are not new to me, but the change in worldview is." Why, I wonder, after spending several decades in the field? "I'm someone who changes their mind slowly," he says.

In 1840, when a German chemist named Baron Justus von Liebig published a monograph breaking down the three main elements plants need for growth, he also demystified soil fertility, which had long been an enigma. Within a few decades those three elements—nitrogen, phosphorus, and potassium—became the basis for the modern synthetic fertilizer revolution, which permanently changed the practice of farming. Since then, however, we've come to understand that plant health is far more complex, and that the relentless use of synthetic fertilizers can in fact do indelible harm to ecosystems and soil fertility in the long run. New layers of soil complexity have more recently come into focus, involving interspecies relationships between untold numbers of microbes and fungi.

Plant personality could be yet another level of that complexity. Right now, individual variation in plants' responses to pests is a mostly inexplicable phenomenon, much as the basics of soil fertility once were. Understanding that not all plants are the same—and the ways they are different—could give researchers an inroad to understanding plants' distinctive behaviors, and perhaps lead to the development of more resilient agricultural crops.

Respecting that individuality, however, will be a greater challenge. Agricultural researchers have warned of the dangers of monocultures— planting a single genetic variety of crop over large swaths of land—ever since the mid-nineteenth century, when a microbe caused a disease known as potato blight which proved particularly deadly to the Irish

Lumper, a staple food crop in Ireland at the time. The devastation of the potato harvest caused mass hunger and around one million deaths. Still, given the economics of modern agriculture, which values yield above all, many of the world's food staples continue to be grown in vast, undifferentiated fields. The crops tend to be bred for productivity above anything else, often at the cost of other traits, like the ability to defend themselves. As such, huge quantities of pesticides and fertilizer are often needed to sustain them. Are monocultures like these fields also a monoculture of a single type of personality, I wonder?

What might happen inside that field, personality-wise, if more genetic variation were allowed in? The culture of the place might change; more lifestyles might live side by side. Plenty of research has already shown the benefits of biodiversity to the resilience of a farm field or an ecosystem. But diversity of personality might be yet another facet of what makes it all work. A field of multiplicity may flourish precisely in relationship to the many approaches to life it contains. As these initial findings illustrate, neither the meek nor the bold can perpetuate a species on their own.

Chapter 4

Alive
to Feeling

We're just a biological speculation
Sittin' here, vibratin'
And we don't know what we're vibratin' about

—FUNKADELIC, "BIOLOGICAL SPECULATION,"
WRITTEN BY GEORGE CLINTON, MAY 22, 1972

Electricity is a wily force. It itself is not alive, but it is very often the best
sign of life. It's a proxy for aliveness—or it may be aliveness itself. Electricity is entangled in every aspect of our living. It is behind our ability
to move, think, breathe. It doesn't have a pulse, but a pulse has it; or,
rather, electricity is the reason for the pulse at all. What to call something that on its own is not quite alive, but surely not inert either? The
theorist Jane Bennett calls it vibrancy. That appeals to me. Electricity
has its own vibrancy. It makes us happen.

Electricity makes plants happen too, or at least that's what the science is looking like. From a certain perspective, a plant is a sack of
water—or slightly more specifically, a skinlike sack of cells, each inflated

by a coursing watery liquid. (Same with us, by the way.) That arrangement makes plants extremely electrically conductive. Electrical pulses move through the plant body very fast. But could plants be using that electricity to understand and react to the world, like we do? To move, grow, send messages to their distant parts? Whereas most electrical impulses in our body are routed through our brains and spat back out as information, plants have no such recourse. So how in the world would electricity be a means of signaling, making meaning out of inputs, without a brain? Scientists, right now, are racing to answer that question. Several confided to me a suspicion about how it might be possible, an idea that verges on the mystical. Or at the very least, it verges on an entirely new conception of life—which so often starts out sounding like mysticism, doesn't it?

Touch the skin of your own cheek. Feel that touch, both in your finger and its landing site on your face. That feeling was brought to you by electricity, an elaborate chain reaction emanating from the cells on your fingertips, and on your cheek, all the way to your brain and back again. In the human body, electricity works like this: the membrane potential of our cells, when at rest, is ever so slightly negatively charged. Positively charged elements—sodium, magnesium, potassium, and calcium ions—are afloat in the plasma between those cells. These are your electrolytes. When touched, the cells open channels in their membranes and allow these ions to pass through them. Think of the sluice gates in canals that let water in and out.

Suddenly, with the influx of ions, the cell's charge flips from negative to positive. This produces a burst of electricity known as an action potential. That sudden burst triggers the ion gates in the neighboring cell to open too, electrifying that cell in turn. This chain reaction travels fast, sending information via the electric current that the bestirred cells make from your finger (and cheek) to your brain and back again. Almost all our cells are capable of generating electricity. Muscles are electrically active every moment they're contracting and releasing; it's electricity that makes this movement possible. The same holds true for the smooth muscle around our veins, which contracts and releases to keep the blood flowing through our body. Our brains, of course, are fantastically electrical, awakening us to the caress of our own cheek before we have time to wonder what the touch feels like.

But what about when electricity diminishes? When humans are put under general anesthesia, they stop responding to touch. Touching an anesthetized person's body—or cutting them open with a scalpel—will not produce the same flurry of electrical bursts it would under normal circumstances. The drugs interfere with our action potentials. Similarly, when researchers put Venus flytraps under general anesthesia—by placing them in glass cases and suffusing the air inside with diethyl ether— the Venus flytraps stop reacting to touch too. They don't snap closed, no matter how many of their trigger hairs are flicked; when the ether is removed, within fifteen minutes the traps snap closed as normal again.

The same holds true for the species *Mimosa pudica*, more often known by its nickname "sensitive plant." In its normal state, a mimosa will close its fanlike leaves at the slightest touch, folding them neatly like venetian blinds. Keep touching it, and the whole leaf will go abruptly limp where it meets the stem, like a wrist. This has a purpose: if you are a caterpillar eating a leaf that suddenly droops, you might fall off. But when mimosa is etherized, the plant won't close its leaves, no matter how much it is touched.

Pea seedlings, which normally wave their tendrils around enough over the course of twenty or so minutes that they appear to be dancing, will curl their tendrils inward and their swaying will grind to a halt when under the influence of diethyl ether. When the ether is removed, they recover, and begin waving around again.

The mystery of plant electricity calls to mind other mysteries; bodily, human ones. Our own electric brains are made from wiring so circuitous that no map has yet been made of every pathway. It also makes me think of the mystery of how anesthesia works on us, the unknown mechanism by which anesthesia can so blithely turn the switch of our circuitry to "off" without snuffing us out completely. We do know that in the human brain, deep anesthesia appears to change the pattern in which electrical impulses flow. Brain waves decrease, resulting in a general dimming of activity. The flow of information, it seems, slows down or flickers out. In some schools of thought, the presence of consciousness is evident mainly in its inverse—by the ability to be knocked unconscious.

In our brain, electricity travels in waves. Information appears on colorized brain scans as pulses, like a wave traveling between two shores. The complexity and coherence of these waves are what neurologists routinely use to determine brain health and mental state. Christof Koch, a chief scientist at the Allen Institute for Brain Science in Seattle, takes it further. He is a fan of a theory developed by neuroscientist Giulio Tononi, which argues that the complexity and integration of these waves are what actually create in us a coherent feeling of reality, one way we sense our own consciousness. Consciousness, Tononi argues, arises from the richness of this wave pattern. Koch, Tononi, and their colleagues have developed a system to, at least in theory, measure how integrated these waves are; the more integrated—meaning the better organized each discrete region of the brain is, and how well those regions are connected together—the higher degree of consciousness it indicates. Using this formula, he believes there is a potential for consciousness in all living organisms. For him, the difference between life forms is not about consciousness versus nonconsciousness, then, but about degrees and intensities of consciousness. A bug has less consciousness than a person, but a bug to some degree is conscious. It's a gradient. And it all comes down to waves.*

This wave form echoes throughout nature. A wave is a very good way to transmit biological information. Slime mold instructs its own movement by sending wavelike pulses through its body, which is it-

---

* Anthony Trewavas, joined by two colleagues, wrote an argument for using Koch's so-called integrated information theory to investigate plant consciousness. See Pedro Mediano, Anthony Trewavas, and Paco Calvo, "Information and Integration in Plants: Towards a Quantitative Search for Plants Sentience," in "Plant Sentience: Theoretical and Empirical Issues," ed. Vicente Raja and Segundo-Ortín Miguel, special issue, *Journal of Consciousness Studies* 28, no. 1–2 (2021): 80–105.

self a single giant cell packed with many thousands of nuclei. Once one leading edge of the slime mold has picked up the scent of nearby sugars and proteins, it softens the nearest part of its gelatinous form, causing the fluid in its body to bulge forward in that direction. The rebalancing of fluid causes the entire sack of the giant cell, numerous nuclei and all, to ripple in a wave, propelling its gelatinous body forward in the direction of the food. Likewise the slime mold can pulse with tiny contractions, sending waves through its fluid body to rapidly send signals to far-flung parts of itself, making its coordinated behavior possible. Fungi, meanwhile, also use waves to consolidate information about their environment into bodily action. Mycelium, the ubiquitous underground body of fungi, may coordinate its millions of individual threads by way of waves of electricity. In this way, information about moisture and food can travel throughout a mycelium whose hairlike tentacles might form a mat spanning a hectare of forest floor. In the case of both the slime mold and the fungus, information is received, absorbed, and translated into a cohesive action without the need for a brain. And often that cycle begins with touch.

Scientists have long observed that virtually all plants are highly sensitive to touch of any kind, and will change their growth accordingly. They even have a word for this phenomenon: *thigmomorphogenesis*. Darwin described touch sensitivity in plants in the late 1800s, but the phenomenon has been known to farmers for much longer. In traditional agricultural practices from many regions, whipping, prodding, or otherwise flagellating certain crop plants was thought to induce heartier growth, or help prevent a plague of pests. In the 1970s and '80s, a plant physiologist in Ohio more or less confirmed this folk knowledge by stroking the stems of plants in a greenhouse each day. Mordecai Jaffe, or "Mark" to most people, found that repeatedly pestering plants made them tougher. He began his investigation by fastidiously stroking several varieties of rather ordinary plants: barley, cucumber, common bean, castor bean, and English mandrake. If he stroked a plant once, it wouldn't change. But if he stroked them over and over, for about ten seconds once or twice a day, they would change

quite a lot. The response was fast: within three minutes of his beginning to rub its stem, the plant would slow or even cease elongating,
which it was otherwise doing all the time. When Jaffe stopped stroking
the plant, it would begin to elongate rapidly, even faster than its normal growth rate, as if making up for lost time. In Cherokee wax bean
plants, the stroked stems would grow girthier, and harden. It becomes
impossible not to make jokes about this, but it was also serious business: Jaffe coined the word "thigmomorphogenesis," and a whole new
field of plant touch studies was born.

Jaffe found the same was true of young Fraser firs and loblolly pines.
Instead of growing tall, the trees would begin to grow stouter and
harden. Jaffe speculated that this response was probably "designed to
protect plants from the stresses produced by high winds and moving
animals." If you're being bumped and bent all the time, it's probably a
good idea to bulk up. The Cherokee wax beans, meanwhile, seemed to
have yet another strategy: becoming elastic. Jaffe decided to see what
would happen if he bent them a bit. It turned out that while unbothered beans would bend a little and then snap, Jaffe's stroked beans
could fold nearly 90 degrees without breaking. So now he knew that
touching a plant could make it shorter, squatter, and more flexible—all
incredibly useful ways to avoid getting killed in a world full of wind
and inconsiderate animals.

Later, the genomics revolution made it possible to see just how im-
pactful touch is to plants on a deeper level. Peering at the genes of *Arabidopsis thaliana*, a weedy plant in the mustard family and the lab rat
of the plant biology world, researchers saw that touch quietly triggered
such a dramatic response in their hormones and gene expression that it
could substantially inhibit their growth. They stroked the arabidopsis
with soft paintbrushes, and then analyzed the plants' genetic responses.
Within thirty minutes of being touched, 10 percent of the plant's genome
was altered. Clearly, the plant was reorganizing its priorities to deal with
the disturbance, and rerouting energy away from the hard work of getting taller. Touched multiple times, arabidopsis cut its upward growth
rate by as much as 30 percent, just as Jaffe had found years before.

When touched, a plant will essentially activate its immune system. In this way, human touch has been shown to help plants ward off a future fungal infection, because the plants' defenses are already up. Whatever the situation, touch a plant, and it will take note, most often by becoming incredibly stressed and defensive. Most plants don't appear bothered when we step on them or pluck a flower. But we now know they bristle internally with all the force of a startled porcupine or a spooked stallion. Plants are fully aware of our contact with them, and will rearrange their lives to respond to such treatment.

But how is this feeling possible? How is the touch noted by a plant, and how could it possibly be translated into a response? The answer might involve electricity. Touch a plant, or an animal, and its response will show up on a voltmeter.

One of the earliest attempts to study electricity in plants was made in the 1900s in Kolkata, India, by a biologist, physicist, botanist, and science fiction writer named Jagadish Chandra Bose. J. C. Bose, as he was known, was the pioneer of wireless telecommunication, having discovered millimeter-length electromagnetic waves—the microwaves that made the first radios possible and are used in remote sensing and airport security scanners today. In fact, he built the radio wave receiver used by Guglielmo Marconi to make the first working radio. He was perhaps the most famous biologist of his generation; he was knighted, elected to the Royal Society, and was the first Indian to hold a United States patent. And yet, outside of South Asia, he has been largely forgotten.

In the years following his major breakthroughs with microwaves, thinking there might be a sort of electrical life to everything, Bose began experiments on vegetables. He attached electric probes to various vegetables, and claimed to record a "death spasm" in the form of a spike in electrical activity. He hooked a cabbage to a voltmeter in front of the playwright George Bernard Shaw, who was reportedly horrified to witness the electrical "convulsion" of the cabbage as it was dropped in boiling water. Shaw, it must be said, was a vegetarian.

Bose also observed how mimosas produced an electric impulse just before their leaflets closed. English scientist John Burdon-Sanderson had first recorded "electrical excitations" in another sensitive plant, the Venus flytrap, in 1876. Yet he was only looking at the surface of the leaf. Bose went deeper, looking at the electrical response within individual plant cells with a microelectrode recording system he designed himself, several years before scientists took the first microelectrode readings of single neurons in animals. He watched as the voltage in individual plant cells changed when they were irritated, clearly responding to touch. In 1925, some years later, he wrote about "plant-nerves," and suggested they behaved like synapses. By then, the earliest explanations of the animal nervous system were being published, though the word *neuron* had not yet been coined.

Plants, Bose decided, must have nervous systems. He was convinced that electrical impulses were responsible for controlling most plant functions, like growth, photosynthesis, movement, and responses to whatever the environment threw their way—light, heat, exposure to toxins. "The results of the investigations which I have carried out for the last quarter of a century establish the generalization that the physiological mechanism of the plant is identical with that of the animal," Bose wrote.

Now, this isn't entirely true—plant cells are different from animal cells; they have cell walls and things like chloroplasts. Plus plants simply don't have synapses. But Bose called it a "generalization," and if we truly are generalizing, he seems right. Plant and animal bodies may be operating on similar basic principles, at least electrically speaking.

I was hardly the first person to stumble back into Bose's plant experiments. *The Secret Life of Plants* had an entire chapter devoted to Bose, which was one of the few parts of that book that later held up to scrutiny. The year the book was released, in 1973, a young biology student named Elizabeth Van Volkenburgh was fresh out of her undergraduate degree. She read *The Secret Life* on her breaks from her job as a technician in a botany lab at Duke University in North Carolina. The Bose chapter stood out to her. Soon the idea of plant electricity came to dominate her thoughts.

I'd first come across Van Volkenburgh's name listed as the president of the Society for Plant Neurobiology—now toned down to the Society of Plant Signaling and Behavior—and saw that she'd studied the electrical impulses in sunflowers years ago. When I called her in 2018, she sounded surprised. Now, as a professor at the University of Washington, she couldn't get her mostly premedical students taking a required ecology elective to care all that much about plants, let alone about early research into why they had electrical currents running through them. She was leading a lab that studied how leaves expand. But I was calling to talk about plant electricity—a preoccupation of hers from years ago, back when she still published papers on this stuff. Back before the funding dried up.

Van Volkenburgh remembers 1973 vividly. She'd just graduated with a degree in plant biology; she'd chosen biology because her grades in everything else were worse. Her work in the Duke lab was mindless; it was her job to endlessly count leaves on experimental plants and measure their length and width ad nauseam. She wasn't sure what she wanted to do with her degree, but this wasn't it. On her breaks she read *The Secret Life of Plants*.

So—she read—plants had electrical lives. Why hadn't this ever come up in her undergrad lectures? For one thing, Bose had had an embarrassing period. He dedicated a portion of his career to questioning whether machines were alive; when his scientific instruments began to slow down after repeated use, he saw a parallel to fatigue in human nerves. The episode reminds me of Alexander Graham Bell, who invented one of the most important pieces of technology in the modern world, the telephone, but who was driven to do so by his belief that the static he could hear on the line was messages from dead people—and might be coming from his dead brother.

Bell wasn't expunged from the canon for that. Neither was Thomas Edison, whose lesser-known ideas included a belief in telepathy. Those parts of their biographies have simply slipped to the background. Of course, they were white men. Bose was dark-skinned and Indian. One botanist told me his lost legacy is the product of point-blank American racism.

In 1981, after Van Volkenburgh got her PhD and was working as a postdoc at the University of Illinois, she began to experiment with plant electricity. She'd arrived to work on a completely different problem, using corn plants. But her advisor had worked with action potentials and showed her how to measure them. She cut off a piece of corn leaf and hooked it up to a voltmeter that beeped when current ran through it. Then she shone a light on the corn leaf. Its cells, still alive, could still photosynthesize; photosynthesis is an inherently electrical process. The voltmeter went wild, beeping frantically.

"I was excited. There's something very elusive about electricity," she said. Electricity is invisible, but stick some probes in a plant, and suddenly you get a signal on a screen. "It was like, wow. It's almost like it's speaking to you. It feels alive."

By 1983 she was back at the University of Washington, where, in another building on the same campus at the same time, David Rhoades had just published his infamous experiments on the caterpillar invasion of the university forest. The news of talking trees was getting around.* Plant communication? In the lab hallways, Van Volkenburgh and her colleagues wondered aloud if it could be real. If plants could communicate through airborne cues, could they do the same with electrical impulses?

We know our own bodies are essentially electrical. It is often forgotten that our present understanding of how electricity governs human nerves and muscles began in plants. The researchers Alan Lloyd Hodgkin, Andrew Fielding Huxley, and John Carew Eccles won a Nobel Prize for figuring out the electrical nature of human neurons in the 1950s. Their work was based on earlier studies, in which scientists measured electrical impulses in the giant cells of *Chara algae*, a common

---

* Seven months later, another paper finding virtually the same thing in maples appeared in the journal *Nature*. Publication in *Nature* is a big deal; it confers weight. Shortly after, yet another paper found that the messages could be passed between different plant species: wafts of chemicals released by a damaged sagebrush could prompt a nearby tomato plant to boost its defenses.

pond weed. The *Chara* cells were enormous, as cells go—ten centimeters long, and a millimeter in diameter—and thus conveniently visible to the naked eye. You could jab an electrode right into one. And they were excitable in much the same way as human cells were.

It took a long time for science to begin to ask plants more electrical questions. In 1992, a group of researchers from the UK and New Zealand found that they could block the chemical signaling in tomato seedlings, but that the plants would still accumulate the defensive proteins when another part of the plant was wounded. But they also noted that when a seedling was intentionally wounded, a flurry of electrical activity was detectable. Could the defensive signal be sent by way of electrical impulses, they wondered, instead of chemical ones?

In a letter published in *Nature*, the researchers even went so far as to say that the electrical activity had "similarities to the epithelial conduction system used to transmit a stimulus in the defense responses of some lower animals." In epithelial conduction, an electric signal is transported cell-to-cell, by way of narrow channels that let ions pass between adjacent cells. "Although plants lack any structures comparable to animal nerves," they wrote, cells in a plant tissue are linked by narrow threads that have "electrical conductivities nearly identical" to those in animal tissues. Could the signal to beef up defenses be passed this way? What would it mean?

Their results provided the first definite evidence for a link between an electrical signal and a biochemical response in plants. Around the same time, Van Volkenburgh felt she was closing in on something significant. First, she studied how cells expanded, and how that led to growth in leaves. Then she published papers on how the outer membrane of cells reacted to different wavelengths of light, and the way in which that changed how a plant grew. The cell membrane, she thought, had a lot more going on than what she was taught in textbooks. In animals, the cell membrane is what governs the flow of electricity.

By 1993, twenty years after Van Volkenburgh began her graduate studies, another scientist finally realized what was going on with the cell membranes of plants. Botanist Barbara Pickard had been working on electricity in plants since the 1970s, and was known to rely as much on her intuition as her data, to the chagrin of her fellow researchers. But she discovered a channel that led straight through the membrane, with its own little gate; it was there to allow the flow of electric current—essentially, calcium ions—through the cells when something mechanically pushed on them; that is, when they were physically touched. Pickard and her team found the first definitive evidence of mechanosensitive ion channels in plants. For the first time, researchers had a way to look at how, at the cellular level, plants experience touch as a physical force from the inside. "Back when I got here, no one believed plants had ion channels," Van Volkenburgh said. "Voltage-activated ion channels are the basis of nerves."

The ions that caused action potentials in plants weren't the same as the ones in animal nerves, and the proteins that regulate them weren't the same either. But still, Van Volkenburgh thought, "you have to wonder if they have nervelike functions." It was impossible to ignore the parallels between the two structures. If plants had nervelike functions, that opened up a world of possibility, and new questions: Could they be said to feel?

The research from the UK and Australia two years earlier had circled around the same idea, but hadn't quite touched it. Now there was proof of ion channels. This significant finding should have marked a turning point toward a lustrous career in what would have amounted to a brand-new field of research. But at that point, plant behavior was emaciated of funding yet again. In 1995, then-president Bill Clinton got word that the U. S. Department of Agriculture was funding studies on "stress in plants" with taxpayer money. He even made a jibe about it in that year's State of the Union Address, implying that he thought the study was about plants needing psychotherapy, and promised to cut such wasteful spending. This general attitude morphed into added skepticism for researchers trying to push the envelope on plant physiology. Funding became harder to get. Pickard, who already raised the hackles of her colleagues by openly speaking her mind about deficiencies in other people's work, further set herself apart by refusing to play by the rules of grant writing. "People felt that she overspoke," Van Volkenburgh says. "But she was way ahead of the game." She stopped publishing, was slowly ostracized by the field, and was forced to give up her lab; she spent the last decade of her career conducting her research out of someone else's lab.

Meanwhile, Van Volkenburgh found that the recent onset of the genetics revolution had made it impossible to get funding for her work on electrical responses in plants. "Everything shifted to genetics," she said. Genes were in, electrophysiology was out. It was difficult and often fickle work; minuscule cell membranes were quite literally touchy study subjects. Funders preferred the clear-cut nature of finding patterns in genetic codes. Plus there was the old resistance to the notion that plants could possibly be that responsive. "Electrical signaling was not accepted as something plants did. I got tired of going up against people's skepticism of the work." Whatever she tried, she still didn't get money. Eventually she gave up applying to grants and shifted her focus elsewhere, away from electricity and toward teaching. In the lab, she went back to studying how leaves grow, a crucial but less flashy botanical mystery. She remained vigilant about new developments in the electricity field, and became a sort of go-between, a rhizomatic runner, connecting the dots between labs and mediating their debates in the background.

Thirty years later, plant electricity is now blooming into a substantial field of its own, spurred on by improved tools and the slow fade of a now tired taboo, a relic of a more paranoid time. Scientists are resurrecting some of that early electricity research from the days of J. C. Bose, but doing it with better tools. Technology has evolved so dramatically that, with minimal investment, anyone can observe electricity in plants at home. You only need an electrode and something with which to read its output. If you attach the electrode to your wrist, a steady line of spikes and swoops will appear. If you attach that same electrode to your houseplant's leaf, and touch it in any way, a spike and swoop will appear on the readout that looks remarkably similar. These are the action potentials—little bursts of electricity—produced, in your case, by the neurons in your heart that are firing at regular intervals to make it pump blood, and in the plant's case—well, no one yet quite knows why they're there, or what for.

The one exception to this sense of mystery is the Venus flytrap, the subject of some of the earliest plant electricity experiments. The plant is famous for the almost animal impression it gives when it closes its trap, which had just been hanging open toothily, quite like a mouth, a moment before (in reality, the trap is a leaf with a hinge). It eats what we can identify as "real food"—insects, like flies—in addition to having the nearly magical plant habit of photosynthesis. It is a delight to watch one of these leaf-maws spring closed, confirming its carnivorous prowess—how has a plant, in a great reversal of the usual fortunes, fatally outwitted an animal? Of course, this happens all the time in slower ways—one need only remember the starving caterpillars, poisoned slowly by leaves in revolt—but we mammals are temporally biased in one direction. We love a quick kill.

The inside of each trap bristles with a few flexible spike-like hairs. Insects, lured by a sugary aroma, brush against the hairs while looking for nectar. In 2016 researchers discovered that these hairs are mechanosensory switches that elicit action potentials, and that the flytrap can actually count how many action potentials have been triggered; an electrical burst registers on a voltmeter when a hair is touched, and the maw snaps shut. Just to be sure, the researchers zapped the flytraps with doses of electricity, not touching their hairs at all. The flytraps closed just the same. This is the clearest example we have of touch-sense in plants where we know for sure that electricity is causing the response.

In all other plants (and within all other parts of a Venus flytrap) big mysteries remain. How does an electrical signal initiated in one place on a plant cause a change in another place entirely? And how, without a brain, does that signal get translated into action? Some sort of internal organization must be taking place to make an electrical jolt in one part lead to a change in another. Between the discovery of sensory switches and nervelike structures, things are adding up to look rather sophisticated in ways previously unimaginable for plants. But still, scientists need a way to make it all fit together.

In a darkened microscopy room in Madison, Wisconsin, a botany professor has begun to draw a map. Simon Gilroy has been thinking about plant electricity for a long time. In 2013, he and his colleague Masatsugu Toyota became the first people to witness electricity moving through a plant body in real time. To their delight, they saw that it moved in a wave.

The first time I met Simon Gilroy he was wearing a bright-blue Hawaiian shirt covered in green philodendron leaves. Botanists love themed shirts. His bright-white hair was parted down the center and lain neatly over each shoulder, nearly reaching his waist.

Gilroy, who is British and given to small jokes, studied in the 1980s at Edinburgh University under Anthony Trewavas, a renowned plant physiologist. The pair were convinced for decades that electricity moved in a wave pattern across the plant body. They both felt it made sense—information moved in a wave in so many other forms of life. They just didn't yet have the technology to prove it.

In recent years, Trewavas has turned decidedly toward using provocative language to talk about plants, aligning himself in the process with a group of botanists who called themselves plant neurobiologists, publishing papers and books laying out scientific arguments in favor of plant intelligence and consciousness. Gilroy himself is more circumspect, unwilling to talk about either of those things, but the two still work together. Most recently, they had been developing a theory of agency for plants. Gilroy was quick to remind me that he was talking strictly about *biological* agency, not implying intention in a thoughts-and-feelings sense. I nodded, and he continued. "Plants, in the time-frame we think of animals operating, do very similar things to animals as far as information processing. They make very complicated calculations about the world around them. It would be incredibly impressive if a human being did that sort of information processing and came to the output that the plants come to." Plants make their lives work in the environment they find themselves in. That, for him, is proof of their agency. Still, the proof is through inference rather than understanding the mechanics. "When you get down to the machinery that allows those calculations to occur, we don't have the luxury of going, ah, it's neurons in the brain," Gilroy said. "The question is, where is the information processing?" Gilroy's work was beginning to allow us to watch it happen, "But at the moment we don't know *how* it works." Observation and understanding often start out very far apart.

When Gilroy is not in his lab, he is teaching the university's introductory biology course to over nine hundred undergraduates per semester. The class covers all the basics, but with a distinctly plant-forward flair. When he gets to the Great Oxygenation Event—the long period in which the earth's atmosphere transitioned away from being a suffocating cage of carbon dioxide to an oxygen-dominated haven—he makes sure one crucial detail sinks in: plants did that. They made the terrestrial world a habitable place for other forms of life to arise, and eventually to be able to breathe. Without them, animal life as we know it would not even have had the faintest shot at clambering onto the evolutionary treadmill. Our cells would never have formed. "Things like mitochondria wouldn't work in the ancestral environment."

The basic view of Darwinian evolution is this: a living organism goes through a wide array of random mutations, until something works, and then it keeps it. That imagines a rather passive view of how life forms. But plants certainly had a hand in their own evolution, and the evolution of the environment. That, for Gilroy, seems to be the central point: plants have designed the world around them to suit their needs. Why don't we get that? We wouldn't be here were it not for them. The idea that they lack agency is absurd, once you have that awareness.

Unlocking certain mysteries would help us understand how plants process so much information so expertly. Gilroy leads a plant science lab that, among other things, regularly sends baby plants to the International Space Station and trains astronauts on how to tend the seedlings so he can study the impact of microgravity on their roots. How plants perceive gravity is a persistent mystery in botany. No one quite knows how they do it. But in humans and many other animals, the way we perceive gravity is understood: in our inner ear, we have canals angled at 90 degrees to each other. The canals are lined with trigger hairs, much like those inside Venus flytraps. The canals are also full of liquid in which crystals are suspended, like glitter in a snow globe. As we bend or turn, those crystals fall down with gravity, settling onto some of the trigger hairs. The hairs bend under their weight like a pin struck by a pinball, sending electrical signals to our brain, which tells us which direction is down. (If you spin around and stop, and the world still seems to be moving, it's because the fluid in those canals is still moving, as if the snow globe was given a good shake. The pinballs are hitting all the wrong pins. The spinning will stop when the ear confetti settles again.) But the key here is that the electrical signal is sent to our brain. Only then does that information turn into something our bodies understand.

"It's a beautiful machine, and we know how the machine works," Gilroy says of our inner ear. Plants have a very similar system: scientists have found falling granules within their cells, just like the ones in our inner ear.* "But then we don't know what happens. There are no hairs, there are no systems to tell you the machines that do the measurement." After the crystal falls, no one knows what happens next. What gets triggered? And where does the signal from the trigger go? Is it transmitted through electrical impulses? It's still a black box. Without a trigger, the mechanics of how the plant senses those falling granules are a mystery. And without a brain, one might expect the information to ricochet around the plant, never arriving at any decision center that can make heads or tails of it, so to speak.

Despite this, the plant is clearly processing the up-down information to decide how to grow—plant roots in general grow down, and shoots, in general, grow up. If you tip a plant over, it will eventually start growing upward again. They are clearly sensing gravity. Plus, they're integrating that information into the information they've already gathered

---

* But made out of starch, instead of ours, which are made out of calcium.

from various other aspects of their immediate surroundings—obstacles, neighbors, the direction of the light, the temperature of the soil. But how? So far no one knows. "And not for want of trying," Gilroy says. "There are really very, very clever researchers who have done what you think would solve it—these are very clever experiments. But we have never found it."

That is quite literally the essence of the entire question of plant intelligence: How does something without a brain coordinate a response to any stimuli at all? How does information about the world get integrated, triaged by importance, and translated into action that benefits the plant? How can the plant sense its world at all, without a centralized place to parse all that information?

A few years ago, Gilroy and Toyota thought they'd have a go at that question. Toyota thought that if there was an electrical trigger associated with gravity sensing, like the kind found in animal ears, it would probably be accompanied by a burst of calcium. Calcium is not, in itself, a form of information. It's basically the footprint left behind by electricity, a kind of "second messenger." In animals, calcium levels increase in a cell when ion channels open. Ion channels open when electricity is passing through. So calcium shows up in a cell directly after the electricity does.

The technology for visualizing calcium in plant cells was dreamed up years ago. It worked like this: researchers took the gene responsible for making green fluorescent proteins out of a species of jellyfish that glows naturally in dark water, and engineered it to be responsive to calcium. Then they inserted the gene into the chromosome of a plant—the part of the cell responsible for passing genetics on to the next generation. When a gene is inserted into the chromosome, it duplicates itself in every cell of that organism's offspring. That means that every future seed that plant produces will make a baby plant with the capacity to glow green already built into each of its cells. Intriguingly, virtually all organisms have the capacity to run the same bit of jellyfish DNA. "The genetic code in the jellyfish is universal," explained Gilroy. "You can take the code and put it into any other organism you want, and it will work the same." Even people? I pictured a person with a faint green glow coursing through their musculature. Gilroy laughed. "Hypothetically you could do it to people. Ethically you could not."

The jellyfish protein turned out to be a fantastically useful lab tool for watching calcium in motion. By now, people had been improving on these green fluorescent proteins for a generation, altering them to make them glow brighter when activated, and they had recently gotten very good. At the same time, microscopes became available with a large enough field of view to look at a whole plant at once, and a camera sensitive enough to detect even relatively weak fluorescence. It was a matter of technology finally catching up to ideas scientists had wanted to test out for years. "It was just fantastic," Gilroy says.

Gilroy and Toyota thought the fluorescent proteins might be a perfect way to study the mystery of gravity. Perhaps they could watch where the signal went by observing the fluorescent path. But before they tried to apply it to the big gravity question, they thought they should have a control, to make sure the system was working. Something that would easily make calcium move around. "Wounding is sure to cause a calcium signal," Gilroy told Toyota. Scientists already determined that plants will produce an immediate spike of electricity at the place where they've been cut, chewed, or otherwise damaged. So Toyota went down to the microscope to cut some leaves, expecting to see a little flurry of calcium in the same spot as the cut. Within minutes he came running back up to their office. "You've got to come see this," Toyota said. "I think we're going to work on wounding."

A wave of green was moving across the plant from the place where Toyota had cut the leaf. The imprint of the cut radiated outward, until the calcium had coursed through the whole body. The visual was clear, stunning. Anyone could understand it: one way or another, the whole plant was being notified of the wound.

"If you're a plant biologist, you know plants respond in milliseconds. That is not a controversial thing at all. You absolutely know that if you put a stimulus on a plant, you know the biochemistry is changing instantaneously," Gilroy says. "But to be able to put it in a way that nonbiologists can see it happen is a big output. To remind everyone that all living beings respond very quickly to the world around them. Because if they don't, they aren't going to be living for very long."

They could now see in real time how incredibly sensitive plants were to touch of any kind. After leaving a plant undisturbed for a good long time (even bumping the table a plant sat on could send a shiver of green coursing through it), Toyota took a plastic pipette tip from the lab and wrote the word *touch* across its leaf. Luminous green waves reverberated outward from the shape of the word. Gilroy later used the video of this moment in the microscope as the final slide in a presentation, right before his contact info, so that the slide read "Keep in TOUCH."

ON A FREEZING day in December, I arrive in Wisconsin to see the green reverberations for myself. I find Gilroy in his office, this time wearing a blazing orange Hawaiian shirt decorated with a surfboard motif. It is minus 12 degrees Fahrenheit outside.

Gilroy takes me into his lab, where Jessica Fernandez, a molecular biologist on his team, brings us a flat of young tobacco and arabidopsis plants she's reared herself, especially for my visit. They are each imbued with the fluorescent jellyfish proteins. Sarah Swanson, the director of the department's microscopy center and the head microscopist in the Gilroy lab, joins us too. Swanson is also Gilroy's wife.

Fernandez lightly drops the tray of plants on the lab bench, and a single leaf on one baby arabidopsis plant catches the lip of a box, bending in half. "Don't stimulate them," says Swanson, wanting to conserve all the plant's reactivity for the microscope. She'd found it was best to startle them from a completely resting state. "It's okay. We'll let them recover," Fernandez says. "And then we'll torture them," adds Swanson.

Swanson leads us into a small room dominated by a microscope attached to a computer monitor. She turns out the lights. Fernandez dips a pair of tweezers into a glutamate solution, and hands them to me. Glutamate is the most important neurotransmitter in our own brains, and research has recently found that it plays a role in plant signaling too, boosting the signal. "Be sure to cross the midrib," Fernandez says, pointing to the thick vein running down the middle of each tiny leaf. If I only pinch the edge of the leaf, without bothering the big veins, the leaf will probably light up in response, but the signal won't travel to its other parts. The veins are the plant's information superhighway. Get the vein, and the pulse will move all over the plant in a wave. I pinch it gingerly the first time, and I can feel the room's disappointment as we wait a few seconds in the dark for the image on the monitor to change. The leaf begins to light up, impressively to me, but I have seen Gilroy's videos, and know it gets better than this. I find it difficult to enthusiastically wound the plant. But Fernandez dips the tweezers and hands them to me again, this time with the instructions to really go at it. I feel like I am in a vegetal version of the Milgram shock experiment. Unwilling to let this room full of scientists down, I pinch harder this time.

The difference is dramatic. The plant lights up like a Christmas tree, the veins blazing like a neon sign. The green luminance moves from the wound site outward, across the rest of the plant, in a bioluminescent ripple. I am watching this plant experience a cascade of feeling. A wave of sensation. As the light travels along the vein system, the image reminds me of something. It looked unmistakably like the branching pattern of human nerves. Swanson whoops. "Oh heck yeah. That's what I'm talking about. That's deluxe." Gilroy hoots. "Save that one." Fernandez claps and saves the video to an archive. Within two minutes, distant parts of the plant have received the signal.

The glutamate on the tweezers, I am told, speeds everything up. The green fluorescence would have shown up without it, but with the addition of the glutamate, the electrical activity seems to become more intense. In 2013 a team discovered that glutamate-like receptors in plants would travel through the plant body, flipping on defense-related genes in plants that had been recently injured. Now, using their fluorescent plants, Gilroy and Toyota have found that adding glutamate makes the glowing green signal move at a rate of about one millimeter per second, lightning quick for a plant. It is much faster than can be explained by simple diffusion, or the passive flow of compounds through the plant's vasculature. It is moving at the speed of electricity.

Gilroy thinks that there likely is a store of glutamate sitting inside each plant cell, and that when a cell is crushed, like by my tweezer pinch, there's a good chance that the glutamate "leaks out," triggering the adjacent cells to "freak out." The punctured cells dump their glutamate, creating bridges between themselves and the other cells, ready for charged calcium ions to cruise straight over. The crushing force of my tweezer pinch had likely created a miniature glutamate tsunami.

It's all a bit like how the animal nervous system works. In fact, Edward Farmer, the researcher who first discovered that genes closely related to the glutamate synapses in our brain are involved in electrical signaling in plants, told me that the first thing he did when he started looking at electrical signals in plants was to buy a neurobiology textbook. Mammals use glutamate receptors to transmit signals quickly throughout their body. Picture a football player catching a pass in the end zone. The football is the glutamate, and the player is the glutamate receptor. Now imagine if the player catching the ball also caused the stadium lights to suddenly become electrified. When glutamate binds to a glutamate receptor, it causes positive ions to flow into the cell, increasing the cell's electrical charge. Whenever you're talking about electrical signaling in cells, you're talking about ions moving across the cell membranes. Electricity in a body always begins with chemistry of this sort. Our synapses, for example, are made up of two nerve cells that communicate over a gap between them, called the synaptic cleft. In this scenario, one of the nerve cells has vesicles packed with glutamate. The nerve cell dumps that glutamate into the synaptic cleft, and that's what triggers the next cell, causing the synapse to fire. It sounds a lot like Gilroy's vision of glutamate dumping in plants.

The presence of neurotransmitters in plants brings up its own intriguing questions. If plants use neurotransmitters to send electrical signals through their bodies, could they be said to have nervous systems? Before I can ask about any potential similarities between human nerves and what's happening in Gilroy's plants, he anticipates my question. "Some of the molecular players might be the same," he says. "The glutamate receptors in plants look like the glutamate receptors in animals." But, he says, "it's not nervous conduction. There are no plant nerves. They don't exist in plants." Still, he concedes that the systems themselves look a lot alike. But you don't have to talk about nerves at all, he says. He prefers "conduits of cells that could allow propagation of an electrical change that a plant uses for information."

Gilroy may not want to call it a nervous system. But, he concedes, it is a striking example of how biology replicates across species. "If biology has something that works well, it pops up in lots of different organisms looking very similar, because why reinvent the wheel when you already have the wheel?"

The lack of plant nerves didn't stop two scientific reviewers writing in a journal that Gilroy and Toyota had found "nervous system-like signaling" in plants. The issue has even leaked out of plant science of late, and people from other scientific disciplines are weighing in. Plants have no neurons or synapses, as far as anyone can tell. And animals have no xylem or phloem, of course. But the way that electricity propagates through the plant to send signals between its various parts has led several scientists to make the comparison, and perhaps none more intriguingly than Rodolfo Llinás, a neuroscientist at New York University, in part because his subject is people, not plants.

In a paper titled "Broadening the Definition of a Nervous System to Better Understand the Evolution of Plants and Animals," Llinás and Sergio Miguel-Tomé, a colleague at the University of Salamanca, basically argue that it makes no sense to define a nervous system as something only animals can have rather than defining it as a physiological system that could be present in other organisms in a different form. Defining it phylogenetically—meaning assigning it only to one portion of the tree of life—ignores the very real force of convergent evolution, where organisms separately evolved similar systems to deal with similar challenges. It happens all the time in evolution; a classic example is wings. Flight evolved separately in birds, bats, and insects, to very similar effect. Eyes are another example; the eye lens has evolved separately several times.

It's reasonable to imagine the nervous system as another case of convergent evolution, Llinás and Miguel-Tomé say. If a variety of nervous systems exist in nature, then what plants have is clearly one. If it walks like a duck and quacks like a duck, it's probably a duck. Why not call that a nervous system already?

I realized that up until the moment in Gilroy's darkened microscope room, I had been struggling to connect everything I was learning about plants to actual plants in front of me. The theory and the physical reality sometimes felt far apart. Or put another way, plants' abilities felt literally incredible. I couldn't credit them to anything I could see. The facts were like radio waves or magnetic poles: I accepted their existence without internalizing their materiality. But watching the green light move through the plant body changed that for me. Suddenly it all became very tangible. I was watching the plant becoming aware, in its own way, of my touch.

I was now several years into thinking about plants, so my sluggishness to grasp this earlier was a bad sign, I supposed, for the fate of news about their prowess as it reached the wider public. What public could be expected to assimilate this information on sight, if it had taken me this long even with such dedicated attention? I realized part of the problem was the way the pieces of information all came wrapped in layers of hedging, language that distances plants from ourselves at all costs. Calling the vasculature of a plant its nervous system could change that. I thought of Theophrastus, and his wisdom that humans needed metaphors they could connect to. The core of a tree should be called the heartwood, he said. No one has ever looked at heartwood and expected to find the vena cava. Yet it works to call to mind the right meaning: here is the tender flesh that keeps the tree alive. And here too are the channels through which electrical signals pulse.

Still, electricity in plants is yet an enigma in one crucial respect: our tissues and organs are all coordinated through electrical impulses too, and we know the end point for all that electricity is our brains. In plants, there is no such perceptible end point. For all we know about the dynamics of sensing in creatures that have brains, the lack of one should mean any electricity that sensing generates ought to ripple meaninglessly through the plant body without producing more than a very localized response. But it doesn't. A plant touched in one place will—we now know, and can watch, through Simon Gilroy's calcium wave videos—experience that stimulus throughout its whole body. As the impact of the touch streams through the plant body in a wave, the plant itself awakens to the realization of the touch, and responds appropriately.

Touch is a tricky thing, from a biology perspective, even in us. The quest to understand how the human body senses touch at the cellular level is still in its adolescence. Major leaps have been made recently; the 2021 Nobel Prize in medicine was given to two researchers who discovered the mechanoreceptors for feeling heat, cold, and touch. But we are still learning how our bodies translate physical inputs into cellular information that can be passed, full of meaning, to our brain. We know ion channels are important to touch sensing in humans, and now we know some of those very same ion channels may be important to how plants sense their world too. We know electrolytes are important for conducting electricity; humans use mostly potassium ions as their electrolytes, and plants use mostly calcium ions. It's still a murky area, but Elizabeth Haswell is one of those scientists whose work has the potential to elucidate it. A biochemist by training, she was fascinated as a postdoctoral researcher by the idea that science still didn't have an answer for how a plant knows up from down—that ever-present mystery of gravity, which may or may not be solved in her lifetime. She went on to lead a lab of seven at Washington University in St. Louis dedicated to finding mechanoreceptors, or the mechanisms through which plants translate physical inputs into cellular information that can be passed, full of meaning, throughout its body. In other words, nothing less than what it takes, mechanically speaking, for plants to understand their world.

Haswell isn't sure where she stands on the plant intelligence debate. "I struggle with having a strong opinion on this topic," she says. "I don't like saying the plant has a brain. I don't like making animals the foundations—they developed differently—we need to approach them differently." Still, something about it nags at her. "I went on sabbatical and thought, I'm going to develop my own view of this. I didn't."

She is working on the most micro of levels: how individual plant cells turn mechanical pressure into chemical responses. Still, she thinks about the bigger picture—the black box. "I suspect that the plant is responding to some of these stimuli at a higher level, organ or whole-plant," she says. She mentions Jaffe's plant-stroking papers, and the way a Venus flytrap will only close if two trigger hairs are tripped within a certain amount of time. "They can count," she said. "If you just touch plants once, they won't make this huge morphological change," but if you touch them repeatedly, they will. "It must be some sort of decision that's integrated throughout the plant. All these inputs have to be inte-grated in some way, but I have no idea how."

Watching Gilroy's calcium stream video made me think of watching films of brain activity, how those light up too. With brains, we have the tools to watch electricity in real time. Something about it looked similar. I thought about Haswell, and Trewavas, and so many others who all seemed to be asking themselves, one way or another: What if it's the whole plant? What if we're looking all wrong? Of course a plant doesn't have a brain—but what if the whole plant itself is something like a brain? I couldn't get this thought out of my head. It was simple, but it seemed to fit. It also seemed possibly very silly.

To my surprise, I found myself asking it aloud one day, to Elizabeth Van Volkenburgh herself, while we sat in the shade of magisterial old trees on the University of Washington campus in Seattle, where she is now a dean. We were talking about action potentials, and where they go, and why the whole plant can respond to something happening in only one distant part of it. "Could the whole plant be something like a brain?" I asked. She smiled. We'd been talking for nearly three hours. She had fifteen minutes left. I'd saved the question for the last moment, in case she was entirely put off by it and ended our talk right then and there. And now I'd done it, and watched her smile, thinking I'd made a fool of myself.

Then she leaned in a little and dropped her voice to a whisper. "I think you're right," she said. "I just don't talk about it."

Chapter 5

An
Ear to the
Ground

It is nighttime in the rain forest of southeastern Cuba, and a long-tongued bat is sailing between the trees, plucking a clear path through the dense canopy at high speed and in total darkness. All gossamer wing membrane and stealth fluff, its whole body weighs hardly a third of an ounce. A paper airplane. The bat lets out a pulse of tones and listens for the echo they return to its oversize jackal ears. A cavalcade of clicks conjure a landscape of objects and air as the tiny mammal tilts its wings to slice between a tangle of vines.

All of a sudden a tone comes back clear and crisp—again and again in the same way, despite the angle of incidence that changes over and over as the bat flies around and comes closer. The sound is irresistible in its clarity, a lure, a beacon in the night. The bat arrives to find a vine hung with a ring of sumptuous wine-colored flowers, their pollen-tufted faces tilted down toward red pitchers full of nectar. The bat unfurls its long tongue and squeezes its face between the flower and its pitcher. It begins to lap nectar, hovering in midair as it drinks. Its back takes a dusting of pollen in the process. Just above the ring of flowers grow a series of glossy leaves, conspicuously ob-long and concave, like upright canoes. The deep, rounded shape re-flects the same clear echo from many angles. To a moving bat in the acoustic mess of the forest, a loud, consistent echo coming from the same location stands out. And for a rare bat-pollinated vine scattered here and there across a thickly vegetated landscape, standing out is crucial.

Marcgravia evenia, this ruby-colored sonar reflector, was the sec-ond vine found to be acoustically tailored to correspond with bats; the first was a flowering vine that grows on the edge of rain forests across Central America. That vine, *Mucuna holtonii*, produces many small flowers, and releases its pollen explosively. To reach the nectar, a bat must land on a flower and press its snout into a slit between two winglike petals. The pressure of the push causes a second pair of fused petals within, called a keel, to burst open. Inside the keel, bent under tremendous tension, is a stamen full of pollen. Once the bat bursts the keel, the stamen catapults most of the pollen load onto the bat's rump.

Scientists watched as bats consistently landed only on the flowers that still had their hidden pollen keels intact, while avoiding the de-pleted ones. There were so many flowers—how did the bats find the right ones? A small concave appendage flanks the unopened flowers, like an extra petal on a hinge; researchers found that this acts as a perfect mirror for the bats' sonar. The echo it sent back from multiple angles was of "astonishingly high amplitude," they wrote, much like the echo from the leaves on the *Marcgravia*. Once the flower had dis-gorged its pollen on a bat butt, the mirror would lower itself, taking itself out of the acoustic arena. Bats could no longer find that flower and were directed instead to the ones with their mirrors still up.

Plants have a particularly close relationship to sound. Sound suf-fuses everything in their environment, so it would make sense for them to take an active part in so vast and varied a sensory world, especially since so many of the creatures the plant must attract and repel make very telltale sounds of their own. In response plants have morphed their bodies to engage with the world of frequency and vibration. It is not an exaggeration to say they've grown ears.

IN 2011, TWO researchers in Missouri did a wild thing: they put guitar pickups on a plant and proved it could hear.

The idea, as with many good ideas, came accidentally. Rex Cocroft, an animal communication expert, was studying treehoppers. These insects are fantastically peculiar looking, with iridescent exoskeletons and, in some species, a single, preposterously tall horn swooping directly up from their head, like a unicorn partial to right angles. Treehoppers, Cocroft observed, intentionally jiggle their abdomens very fast, sending a thrumming vibration down through their legs and into the branch of the tree or woody shrub they stand on. Those vibrations travel through the plant and are picked up by other treehoppers, who are equipped with highly sensitive legs adapted to the task of acting like phonograph needles. It was, Cocroft had found, a treehopper way of saying "Hi, I'm here." The insects were essentially using the plant as a tin can telephone. It was interesting work, but one day every recording Cocroft tried to make of these vibrations was contaminated by another noise. It was grating. It was rhythmic. It wasn't a treehopper. "There were all these caterpillars chewing," said Heidi Appel, a senior research scientist at the University of Toledo, Ohio, who became his collaborator. An alluring possibility dawned on her.

Caterpillars are the can opener machines of the insect world. "Actually, I like the sound," Appel said when I suggested the analogy. Amplified to volume a human can detect, caterpillar chewing sounds like blocky goat teeth masticating dry hay, or a handful of gritty pebbles being rubbed together in one's hand. Which I suppose can be a strangely satisfying sound, like a cartoon character chewing a carrot. But without amplification, it's incredibly subtle; the sound of caterpillar chewing vibrates a leaf up and down by only a few ten thousandths of an inch.

Appel met Cocroft at a seminar at her university, during the cookies and coffee break. The two introduced themselves by way of what system they studied, a social behavior typical of natural scientists. "I work on how plants can tell they've been damaged and what they can do about it," Appel remembers saying.

"I work on how animals communicate to each other through vibrations in plants," Cocroft said. He told her about the problem with his setup he'd run into a few days before. "It wasn't working 'cause there was a caterpillar feeding on it," he said.

There was a pause in the conversation. They looked at each other.

"You don't suppose the plant is *using* those?" Appel said.

"It was just an aha moment," she remembers. Together they hatched a series of experiments. The reasoning went something like this: caterpillar chewing is ubiquitous in a plant's life. It makes a very distinctive sound. Acoustic vibrations travel faster along a plant body than virtually any other signal a plant could possibly pick up on. Wouldn't it be advantageous, they thought, for the plant to be able to sense them?

But they were wading into troublesome territory. The specter of *The Secret Life of Plants* still loomed over botany, some forty years after publication. Asking whether plants could have evolved to hear—or at least interpret vibrations that we think of as sound—was sure to raise eyebrows. Even Appel's husband, fellow plant scientist Jack Schultz, was a detractor. Schultz had been one of the earliest to claim that trees communicated through airborne chemicals, a very popular thing to wag one's finger at in the 1980s, if you were a plant scientist. It took years, until at least the mid-2000s, for chemical communication to be seen not as ridiculous but as a scientific fact. "He looked at me and said, 'You're nuts. You're crazy,' " Appel remembers. "That's the inherent skepticism of science," she says, charitably. It was a warm day in early September, and she was standing outside her house near Toledo, Ohio, looking at a tree. Schultz was indoors, working on their latest co-authored paper. They'd been collaborating for over thirty years.

Appel isn't a great believer in the plant intelligence debate that's overtaken their field. She'd prefer to relegate it to the philosophy corner while the scientists do the hard science. The words scientists use are important, because what they're working with is complex; using mushy words like *thinking* or *communicating* just confuses things.

"I'm humbled by how much I don't know. But when it comes down to how we define things, I'm not sure that's reconcilable." Still, there is no doubt in her mind that plants can sense sounds.

"Oh my gosh," she said, turning back toward the tree. A huge pale paper wasp nest was hanging from a branch. She stood there and admired it a moment and kept walking around her yard, three acres of majestic oak floodplain forest with a fox run through it. She reached the little sugar water feeder she'd hung out for the hummingbirds. It was empty, depleted faster than seemed possible for hummingbirds. But not too fast for a colony of wasps. "Ah," she said. "What I've done is created the conditions for the wasp house."

Plants and insects interact all day long, and at every stage in both of their life cycles. It may be the most important relationship in either of their lives, if the insect is the type that drinks nectar or eats leaves, which is to say the great majority of them. Plants and insects together make up about half of all multicellular organisms on earth; it wouldn't be an exaggeration to say theirs is one of the most important relationships on the planet. When they decided to test plant hearing, Cocroft and Appel were dealing with cabbage white caterpillars, pudgy grass-green creatures that can devour a leaf quite fast. Here is how a cabbage white caterpillar eats a plant: it sticks each of its nub feet on opposite sides of the paper's edge of a leaf, raises its thumb of a head, and begins to chew in a downward line, back toward its own body. Then it unsticks and resticks each of its nub feet in a ripple back to front, and in this way undulates wormishly forward a hair. And then it does it again, raising its head, chomp-chomping down in a line, and after a few scoots it leaves behind negative space in the shape of a crescent moon where green flesh used to be. Look at any leaf: if you see its edge cut in crescents like a paper snowflake, a caterpillar was there and is now briefly satisfied.

The plant has every interest to avoid this inconvenient dismantling—all those useful chloroplasts, all that photosynthetic potential, carried off in the belly of a gelatinous insect. The good news for plants is that they've built a lot of ingenious ways to put an end to this while a caterpillar is midmeal, or at least to avoid its cousins joining in. As we have seen, some plants will pump out bitter tannins in a bid to taste disgusting. Others will manufacture their own insect repellant, which in many cases is the bit of the plant humans most enjoy—it's the rich oregano oil in oregano, the sharp spice in a horseradish root. Sometimes the approach is more sinister. One devilish case has been found in the humble tomato: the tomato plant will inject something into its leaves that makes the caterpillars look up from their chewing and turn to eye their fellow caterpillars. Soon, the leaf becomes irrelevant. The caterpillars begin to eat each other.

But as we saw in Gilroy's lab, the leaf's reaction to being bitten isn't limited to just that leaf; the bite triggers a cascade of hormonal changes in the whole plant, which means different plant parts have a way of talking to each other. Electricity appears to be one explanation, but even the rate that electricity moves through a plant body—on the scale of 0.05 meters per second—is slower than some of the reactions scientists have observed. One way this threat might be communicated, it seems, is through vibrations we pick up as sound. Acoustic vibrations travel extremely quickly. In a rigid, woody plant, they travel at thousands of centimeters per second, a rate that declines in parallel to the plant's general floppiness but is still, in all cases, very fast. Could plants be said to hear their invaders?

To find out, Appel and Cocroft decided to test arabidopsis with the sound of something that would surely eat it: a cabbage white caterpillar. For the experiment, they decided to use piezos—guitar pickups—tuned to the exact frequency of a cabbage white caterpillar chewing. As a control, they clipped piezos to a group of other arabidopsis plants, but those were left silent.

In their first experiment, they played back the sound of the caterpillar chewing, sending minuscule vibrations through the leaves. But how to test if the plant was responding? "A plant attacked can respond right then, or it can take note of what happened and be primed to react more quickly later," Appel said. So they took the guitar pickups off the plants and tested them with actual caterpillars. Then they had to wait to get the leaves analyzed in a lab, to see if they really were producing defensive compounds.

"Really?" Appel said aloud to an empty room when she saw the results. She went to the lab and asked the technician to please double-check her numbers. The technician sent them back again. Still bonkers. The signal was clear. The plants could hear the caterpillars. She called Cocroft. "You're not going to believe this." Then they got together and tried to think of every possible way they could have made a mistake.

"Maybe this is plants responding to anything, not specific to insects," she surmised. They repeated the experiment with lots of controls. They used a small fan to simulate a gentle wind; maybe that made the plants put their defenses up? They tried playing back the sound of a leafhopper mating song; it has the exact same amplitude as the sound of caterpillars chewing, but it forms a different rhythmic pattern. The arabidopsis did nothing in response. Leafhoppers, after all, don't eat arabidopsis.

All of this work only made it clearer: the plant was responding specifically and exclusively to the sound of its genuine predator chewing. "Of course that resulted in big grins," Appel said. "In science, progress in understanding things is largely incremental, and most of us spend our careers . . . let's just say it's very common for experiments not to work. But when they do, what they tell us about how the world works, they tell us in tiny pieces. It's like bricks in a wall. They accrue." But this wasn't a tiny piece. It was evidence that plants could really hear, in their own earless way. Sound, to them, is pure vibration. And they will do something about it when they sense a vibration that they know is associated with their own harm. Like a caterpillar mouth masticating plant flesh.

Once one opens the door to plants registering the sounds of a caterpillar chewing, other considerations come into play. The world is a noisy place. What else might plants hear?

As I write this, researchers are busy birthing a field some have taken to calling phytoacoustics. That plants should hear becomes more believable when one tries to take a plant's point of view. Hearing is an extremely useful sense to have, especially if you are rooted in place. If you can't flee or seek, at least not very fast, then you need all the advance warning you can get. At an even more basic level, hearing is a sense, ancient and everywhere, fundamental to life. Plants have a lot to gain from using acoustic information. If something is happening outside an organism that could be useful for its survival, that organism may have developed a way to sense it. Evolution, ever scanning for a benefit, will give the organism ways to use its awareness to further its project of survival.

And it could be wildly useful to agriculture, if scientists can find the right application. After all, in Appel's work, a sound cue caused the plant to make its own pesticide. If plants could be made to produce pesticides through simply playing sounds to them, it could reduce or eliminate the need for synthetic pesticides on farms, and in some cases increase the levels of compounds that the crop in question is grown for. In a crop like mustard, for example, the plants' own pesticide is the very thing it is farmed for—mustard oil. Putting a lavender bush on high alert by playing the right sounds would cause it to make more of the defensive compounds we prize in lavender oil.

Researchers globally have tried to see if playing certain tones to plants can prompt them to certain actions. They experiment with various frequencies for different lengths of time. Right now, the research on tones is rather scattershot. One study found that playing arabidopsis a series of tones for three hours per day over ten days increased its ability to fight off a harmful fungal infection. Another found that playing some tones to rice for an hour improved the plants' ability to survive drought conditions. And researchers who played tones at different frequencies to alfalfa sprouts for two hours saw that they increased the plants' content of vitamin C, which means their nutritional value went up. When they repeated the experiment with broccoli and radish sprouts, they managed to increase their content of flavonoids, too. One can imagine a future where farmers set up boom boxes instead of crop dusters.

Appel's work fits into this matrix, in a way, but rather than auditioning random tones in front of plants, she is more interested in asking about the sounds a plant actually encounters in nature. She thinks plants are more likely to have interesting responses to sounds they have evolved alongside. Scientists call this "ecological relevance." Sounds of predators are certainly ecologically relevant. If playing the sound of the caterpillar that eats arabidopsis could prime that plant's immune system, there's no reason to think the same wouldn't be true for other plant-predator and plant-pollinator pairs. Some flowers are buzz-pollinated, for example; they can be induced to release their pollen when played a recording of bees buzzing. Could plants also listen for the sound of their fruit eaters, which are often noisy—think parrots—to time their ripening? Or the sound of thunder to prepare to receive the rain? It would make sense; a desert plant would need to ready itself to take up as much water as possible, and any plant with pollen in its flowers would do well to close its petals ahead of a rainstorm, before the pollen is washed away. Phytoacoustics wants to find out.

The next logical question is how plants can possibly hear at all. They may not have ears in the traditional sense, but ears come in all forms. In 2017, a collaboration between researchers in China and the United States found that the tiny hairs on arabidopsis leaves function as acoustic antennae, picking up and vibrating at the frequency of incoming sounds. Many other plants also have tiny hairlike structures on their leaves; understanding whether or not these structures, called trichomes, function as antennae on other species too will require more study. Researchers have already found that trichomes allow plants to sense the footsteps of moths and caterpillars, and mount defenses in response; trichomes are clearly exquisitely sensitive organs. One can't help but be reminded of the inner ears of animals, which are also covered in specialized hair cells that vibrate in response to sound waves, and convert those vibrations into electrical signals that are sent along nerves to the brain. This is yet another reminder that when evolution has a good idea, we'll likely see it across the spectrum of life.

Now research is emerging to suggest that sound may be so vital to the life of plants as to contribute to their shape. In 2019 researchers at

Tel Aviv University found that the beach evening primrose—a lemon-yellow teacup-shaped flower that grows low to the ground—would increase the sweetness of its nectar within three minutes of being exposed to an audio recording of honeybee flight. The primrose would completely ignore sounds that fell outside of the frequency of the hum of bee wings. The team, led by evolutionary biologist Lilach Hadany, theorized that the sweeter nectar—it had a higher sugar content than flowers not exposed to bee sounds—would better entice pollinators, and increase the chance for cross-pollination.

Multiple pollinators are known to congregate around plants that another pollinator has visited a few minutes before. It would make good sense for the plant to anticipate the bee, in that case. But could the teacup really be a satellite dish, listening for its pollinators? Hadany and her coauthor Marine Veits, then a graduate student in Hadany's lab, found that when they played the bee recording again, this time with a movement-tracking laser trained on the evening primrose, the vibrations of the flower matched up with the wavelengths of the bee recording. The flower was acting like an amplifier, its whole form like a sort of resonance speaker. The team then plucked off a few petals, breaking the flower's perfect bowl, and retested the flowers; this time, it was unable to resonate at the bee frequency. The flower, in this case, was definitely the part of the plant responsible for "hearing"—and it suggests that it had taken on the bowl shape for exactly the same reason satellite dishes are concave. "We found a potential hearing organ, which is the flower itself," she said. When she looks at flowers now, she sees ears everywhere.

Roots, it seems, can be just as acoustically sensitive. Why have ears above ground only, when half your body is under the dirt? There's plenty to hear down there too. Just ask a mole. Or, if you're Monica Gagliano, you might prefer to ask a pea.

The pea seedlings in Gagliano's lab at the University of Western Australia looked like they were wearing giant plastic pants. The curly top of each young shoot peeked out from the top of its own PVC tube. Each tube forked at the bottom into two legs, like an upside-down Y. Gagliano was testing the peas' ability to hear, and more specifically, whether they could hear the movement of water. The PVC pants were effectively Y-mazes, the same conceptual structure used to test learning and behavior in lab mice. In this case, the Y was testing which direction the pea seedlings' roots would decide to grow. At the bottom of each pant leg, Gagliano placed a different tray. After some days of growing, the peas' roots would come upon the fork in the pipe, and have to make a decision—much like a mouse deciding where to turn in a maze. In the first set of experiments, one tray contained a few teaspoons of water, and the other was empty. It's a well-observed fact that plant roots can detect "moisture gradients" in soil, allowing them to find water in close range, and as expected, nearly every pea shoot grew roots toward the water tray.

Next, Gagliano redid the experiment, but instead of water freely available in a tray, she pumped water through a sealed plastic pipe near the base of one of the legs of the Y, while the other Y-leg remained above an empty tray. An aquarium pump continuously replenished the water. This time, there was no way for the plant to detect moisture; only the live sound of running water was available as a cue. But again, nearly every pea plant grew its roots toward the sound of the running water. Next, they were given the choice between the tray of water and the water flowing in the sealed pipe. In this case, they chose the open water, suggesting the plants prioritized actual moisture—a guaranteed drink—over the sound of it. The seedlings appeared to be able to parse various sensory cues, ranking the inputs in terms of priority to their own health, Gagliano thought. But more pressingly, they could hear—and move toward—the sound of real running water.

This would likely not surprise a plumber. Plumbers are accustomed to the frustrating phenomenon of tree roots bursting through sealed water pipes. Cities spend millions each year repairing municipal pipes punctured by "root intrusion." Germany, for example, spends an estimated thirty-seven million euros per year repairing root-burst pipes. The U.S. Forest Service points to root intrusion as the cause of more than half of all sewage pipe blockages.

Now Gagliano is urging her fellow researchers to think broadly about what else plants might possibly hear. If plants can hear animals, can they also hear each other? It has long been known that plants emit very quiet clicking noises when air bubbles in their stems pop as water travels up them. This process is called cavitation, and these "cavitation clicks" seem to increase when plants are dealing with drought stress. This makes sense; less water may mean more air bubbles in the stem. Gagliano wondered if the clicking noises might be intentional utterances themselves, and not just bubbles accidentally popping.

Hadany, who was behind the evening primrose study, made a discovery in 2023 that is the first solid evidence that the cavitation click theory could be true. She and Yossi Yovel, who studies bat sounds, put microphones up to wheat, corn, grapevines, and cactus, and recorded their ultrasonic clicks. I listened to the recordings, sped up and amplified to an audible volume. They sounded like popcorn popping, or like a person aggressively typing.

Each species of plant seemed to have their own click frequency. A cactus sounded very different from a grape, for example. But most intriguingly, the clicks dramatically changed according to the plants' condition, Hadany explained. There was a big difference in sounds from a stressed, dehydrated plant and a watered, healthy one. Tomatoes, for example, made thirty-five sounds per hour on average when drought-stressed, versus fewer than one per hour on average when the plant was given all the water it needed. The clicks also increased sharply when she clipped a leaf, acting as a stand-in for a munching herbivore. Unbothered plants were very quiet by comparison. "Tomato and tobacco, when they're feeling well, they emit very few sounds," Hadany said.

Hadany's team developed machine learning models that were capable of distinguishing between plant sounds and general noises, and could identify the condition of the plants—dry, cut, or intact—based solely on the emitted sounds. That certainly opens a possibility that farmers, equipped with ultrasonic sensors, would one day be able to listen for the water needs of their plants.

But more intriguing is what this might mean for plant communi-

cation. Identity and health status: that's a lot of information available
to anyone who can hear it. Humans can't, without amplification. But
moths can. So can bats, and mice. The sounds Hadany recorded would
be audible to small creatures from up to sixteen feet away, according to
her measurements. Could animals—or enticingly, other plants—detect
and interpret those sounds? In other words, could the plants be communicating with sound? "If we can tell, then other organisms can tell,"
she says.

Hadany was cautious on the phone with me not to oversell her
findings; they don't tell us anything about intention on the part of the
clicking plant, she said. Clicking could be just a byproduct of a physical
phenomenon, like our stomachs growling when we're hungry. "I'm not
saying language, yet. Because language assumes both sides." But even
in the most conservative case, she says she considers it likely that someone is listening. If rapid clicks mean the plant is coping with drought,
or besieged by insects, other plants might use the noises as a warning.
Perhaps they close their stomata or raise their immune response. That's
what Hadany plans to study next, and she just secured a major grant
to do it.

But what would that mean for the plant sending out the clicks?
Could they be intentional? That's where it gets tricky. We know that
once another organism starts using information provided by another
living being, evolution often steps in to fine-tune the organism doing
the providing. "It might be entirely passive, but if others are responding to the sounds, then natural selection can act on the emitter," Hadany says. In other words, the sounds could have evolved beyond their
humble origins as accidental noises. They might now be optimized to
serve a very real purpose—like communication. "It's complicated. We
arrived to this prediction thinking about tools for communication. Science is a long process. We're not there yet," Hadany says winkingly.

Gagliano compared the question to bat sonar. Over a century after
the first evidence emerged, science refused to believe that bats could
be using sound to orient themselves in space. It seemed too far outside
assumptions of what animals could possibly do. Scientific disbelief

hampered the discovery of echolocation in bats; could not the same be happening with plants?

Indeed, others have suggested echolocation as one motivation for a plant to make a sound at all. Climbing vines are known to circle the air as seedlings, looking for an upright staff to climb—and seem to locate the position of an appropriate climbing surface long before actually coming into contact with it. Stefano Mancuso, one of the original plant neurobiology advocates and a frequent collaborator of Gagliano's, has used time-lapse video to observe this phenomena in bean plants as they search for, and locate, a nearby metal pole. Again, cavitation—the purely coincidental sound of air bubbles popping as fluid moves through the stem—seems like a logical explanation. But is anything a coincidence in a living body? Mancuso speculates the vines may be using echolocation to sense the position of the pole. Gagliano sees it as following basic evolutionary logic; it's advantageous for plants to learn about the surrounding environment by making sounds, because "acoustic signals propagate rapidly and with minimal energetic or fitness costs," she says. Still, the hard evidence is yet to come.

Could plants have something to say? Gagliano wants us to find out. At this point, these questions have not been answered, nor have they really been asked, in terms of actual experiments. But up until recently the same was true for that other, now-accepted mode of plant communication from earlier: chemical signaling. "The birth of plant chemical ecology, for example, unveiled the strikingly 'talkative' nature of plants and the eloquence of their volatile vocabulary," Gagliano wrote.

Already, scientists have found compelling evidence that language is not entirely confined to the human realm; prairie dogs appear to use adjectives, specific repeated sounds they use to describe the size, shape, color, and speed of predators. Japanese great tits have syntax; they use distinct strings of chirps to instruct their comrades to scan for danger, or tell them to move closer. We've heard about songbirds using backchannels for alarm calls, and risk-averse chipmunks screaming at the slightest spook. Perhaps it would be small-minded of us to foreclose on the possibility of a sound-based plant language emerging too.

To bring up Gagliano's name among botanists is a highly divisive act at this point in time. She has become a contested figure in the field, even as her profile beyond the ivory tower grows. In 2020 a graduate student at University of California–Davis tried to replicate a particularly radical pea-learning study of hers, which put peas through a Y-maze and found they could learn to associate benign cues with rewards, much like animals. In this case, the cue was a gentle wind from a fan, and the reward was light. The conclusion would be world-changing if true: associative learning is a crucial measure of intelligence in animals. But the graduate student couldn't make it work. His peas didn't show signs of learning. I began to hear more murmurs that basically discounted Gagliano all together. The whole episode damaged her reputation within botany circles. Still, it must be noted that replication is a tricky thing. The ability for multiple independent people to repeat a study and get the same result is absolutely crucial for confirming new scientific conclusions, but its impossibility doesn't always mean the original outcome is incorrect. It does, however, mean the study design isn't sturdy enough to hang one's hat on. If the conclusions are in fact true, they'll have to wait for a better experiment to prove them out.

For others, it was Gagliano's 2018 memoir, *Thus Spoke the Plant*, that put her reputation over the edge. In it she describes taking ayahuasca in a shamanic ritual in Peru and communing with the spirit of the plant, who told her how to best design her studies. In science there is a tacit separation of church and state. Purity in science means not wading into mystical waters, or at least, if you do, keeping it to yourself. Science is made up of people, and if those people don't like your work, or if they see you as not one of their tribe, you might get heckled, and you might not get funded. Gagliano has been heckled, at conferences where men (they're always men) have stood up to berate her, and in journals where groups of botanists (again, men) have written letters in protest.

But others have been less harsh. They don't understand why she has been so pilloried, when plenty of men in academia have gone down more mystic routes. Others are more nuanced: sure, it seems like her study design on the pea-learning paper may have been faulty, but her ideas were good, and they were thankful that someone was pushing their field to ask bolder questions, particularly around plant acoustics. Her work advocating for phytoacoustics had made a real difference; it *was* time to take the world of plant hearing seriously.

It is remarkable, meanwhile, to see the gulf between how the scientific and nonscientific world has received Gagliano. She speaks to packed audiences at conferences on philosophy and at science events geared toward the general public. A profile of her ran in the *Style* (not *Science*) section of the *New York Times* in 2019. WNYC's *Radiolab* profiled her work in an episode, and she's given interviews for half a dozen other mainstream outlets. Hers are science ideas that resonate deeply with those outside the institution of science itself.

It remains to be seen how history views Monica Gagliano, but I see her as a perfect symbol of the times. She has a foot in two worlds, just as the emerging science around plants is forcing a confrontation between them. She has written papers in feminist theory journals that seem to advocate for scientists to use more of a felt sense in their methodologies, acknowledging that this is completely antithetical to how they are trained. In a 2022 paper coauthored with University of California–Davis anthropologist Kristi Onzik, she cited Nobel laureate Barbara McClintock's unusual methodology, which ultimately led to her field-shifting breakthrough in 1944 about the nature of corn genetics. McClintock found that some genes could jump, spontaneously changing their position in the chromosome, a previously unheard-of proposition. McClintock would observe her corn plants for hours, "losing herself" in the act of reverence and close listening until she felt she had a "feeling for the organism," and was newly capable of "direct communication" with it. It took years for the molecular-scale technologies to arise which proved to her colleagues that her discovery was true. Perhaps scientists should be more open to losing their grip on the rationalist certainties that sustain their careers, posit Onzik and Gagliano.

Onzik, whose chosen focus as an anthropologist is on the culture of the researchers who study plant behavior, accompanied Gagliano to her lab in Australia, where she was attempting to study whether roots can choose to grow in a path of least resistance by anticipating and avoiding roadblocks. It was similar in design to her experiment on roots sensing the direction of flowing water, but this time she was using more complex mazes, with four arms instead of two. Onzik watched as Gagliano grew frustrated with the sterile conditions in the lab, and the prodigious waste presented by the Styrofoam boxes her plexiglass maze components arrived in. "Without much hesitation," she packed up her mazes and traveled instead to a house in the subtropical forest of New South Wales, Onzik wrote. There, surrounded by "spiders, newts, and snakes," she set up her experiment again. The new environment was not sterile or temperature-controlled. She had now departed, deliberately, from the realm of replicable science. Whatever happened next was unlikely to be admissible in conventional scientific journals. But she was feeling around for a different way of knowing, Onzik writes.

The line between spiritual and scientific worlds is a treacherous position to occupy, and certainly subject to criticism and even contempt from peers. Yet Gagliano seems comfortable there, and unflinching. I see her as attempting to bring those two worlds into relation with each other. In some ways, she exists at the crux of the war over plant intelligence. It probably helps that she is now funded by a million-dollar Templeton World Charity Foundation grant that supports the study of "Diverse Intelligences." She is affiliated with Southern Cross University, in Australia, but her funding is no longer tied to traditional federal sources. Maintaining academic palatability is no longer a concern, at least not for financial reasons.

Gagliano spoke at Dartmouth in early 2020, about humbling ourselves as humans: "We are the new kids on the block. Traditionally, you should pay respect to the elders," by which she meant bacteria, fungi, and plants. She called the view of humanity at the top of an evolutionary chain "arrogant" and "juvenile."

"Who said that science is the only way of knowing? As a scientist, I love science," Gagliano said at Dartmouth. "I think it's a beautiful way to describe the world. But it is not the only way."

I'm struck by the existential conflict that the idea of a hearing plant has kicked up, ostensibly a conflict between science and spirituality. But I'm struck too by the realization that no one is disputing that plants can actually hear. What they are listening for, of course, remains to be fully determined. But it is an earth-shaking revelation on its own for someone who has spent her entire life so far operating under the assumption of the opposite. My world of sound always seemed fundamentally separate from anything plants might be involved in. But lately my idea of what a plant is altogether was changing. They apparently intrude on all levels of our sensory world. Suddenly the sphere in which I exist—the feeling, sharing, hearing world—seemed to be losing its fundamental separateness from the world of foliage. A wall I'd unwittingly constructed between the two was getting thinner, more transparent, like the wet membrane of a soap bubble threatening to tear. Hard green buds were poking through.

Chapter 6

The
(Plant)
Body Keeps
the Score

It's an unseasonably warm day in September, and Berlin is lit by full sun. In this city of long gray winters that begin promptly in autumn, there is a sense that this sunny day might be the last for a long time. People are sprawled throughout the city's parks, lying between hedges and rose bushes. I see three older men on a bench, sitting with their faces tilted upward, eyes closed, saying nothing to each other. They look like they are trying to drink the last of the light through their pores.

The drabness has already overtaken the Berlin Botanic Garden, but a few hardy plants are still in bloom, their flowers angled, too, toward the waning rays. My walking companion, Tilo Henning, is a researcher here. He is telling me about *Nasa poissoniana*, a plant in the flowering Loasaceae family that grows in the Peruvian Andes, and what he is saying has me captivated.

"What do you mean, the flower remembers?" I ask. "Where does it store the memory?"

Henning shakes his head and laughs, his black hair tied in a low ponytail draped over the collar of his sweatshirt. He doesn't know. No one does. But yes, he says, he and his colleague Max Weigend, the president of the botanical garden a few hours away in Bonn, have observed *Nasa poissoniana*'s ability to store and recall information. They discovered that these multicolor starburst-shaped flowers were able to remember the time intervals between bumblebee visits, and anticipate the next time their pollinator was likely to arrive.

This work offers a new and explosive addition to the world of plant behavior: plant memory. I've come here because it occurred to me that memory must be the basis of all complex behavior. I'd learned about plants listening to their surroundings, feeling touch, and exchanging information. But each of these abilities were limited by their fleeting temporality. What good is all that sensation without the ability to remember it? Without memory, very little can be done intelligently. Memories give us the capacity to learn, and to orient ourselves in time and space. What would it mean if a plant could remember? Not the genetic sort of memory, of birds returning to the same migratory grounds each year, but individual memory. Elastic memory. Memories that change when circumstances do.

The elaborate alien-looking flower structures and painful stinging hairs of the Loasaceae family have absorbed Henning and Weigend's attention for decades; they've named dozens of new species and described the nettle-like barbs on the plant's stem, which have menaced them both with plenty of blisters. Weigend is especially enthralled with things that prick and sting; he'd discovered that plants in the Loasaceae family use the same ingredients to make their stinging hairs as humans and animals use to make their teeth. This is advantageous, because stinging is tough work; the hairs are built precisely like hypodermic syringes, and have to be hard enough to pierce the exoskeleton of their enemy and inject them with their stinging toxin. When he looked at other plant families, he saw that the architecture of stinging hairs was remarkably specific to each species, with different combinations of minerals in each, perhaps calibrated to match the hardness needed to pierce the skin of whatever animal ate them. But one year, after an experiment inspired by watching bees flit around their plants in a greenhouse in Bonn, Henning and Weigend realized something new. *Nasa poissoniana* is able to present its pollen when it expects a pollinator to show up. And it does so by remembering the time interval since the pollinator's last visit.

Nasa poissoniana was by then, thanks to their work, already talked about as "the flower that behaves like an animal." Like many plants, the flower carefully parcels out the pollen it presents, only displaying a little at a time, so no single moth or bee gets too much, which would be bad for the overall project of genetic diversity. But *Nasa poissoniana* goes a step further; when it notices fewer pollinators around, it offers up larger globs of the sticky pollen at a time, to hedge the bet that they will only have a few decent shots at pollination. It also dilutes its nectar, to prompt the flying creature to come back twice for the same amount of sugar—dusting its body with pollen both times. Manipulating your pollinator makes sense for a flower living in such harsh conditions. *Nasa poissoniana* thrive at high elevations—between one and three miles above sea level—and often in tiny populations. They have to make every shot count.

Nasa poissoniana is one of very few plants that move their body parts fast enough for a human eye to watch—in this case, they move their stamens from horizontal to vertical within two or three minutes. The flower's stamens start out lying down, each one tucked into one of the concave petals that halo the flower's center like so many canoes. When a bee arrives at the flower, it slips its strawlike mouthpart beneath a central, scallop-shaped petal and lifts up. Beneath the scallop is a pool of nectar; the bee drinks. Somehow, the lifting of that scallop triggers one of the flower's several stamens—the male fertilizing organs of the flower—to, well, erect itself. The mechanics behind that response are still a mystery. But the rising of the stamen is a thrill to watch. Up goes the slender white filament, topped in a little yellow package of strategically parceled-out pollen, to an arrow-straight ninety degrees in the flower's center. When several stamens are erect, they gather in a slender cone shape in the center of the petals, and the flower takes on a striking resemblance to a sci-fi laser-beam launcher.

Other quick-moving plants have clear motivations. A white mulberry, for example, can catapult its pollen at around half the speed of sound, giving it a fair shot at dispersing far enough to find suitable conditions for growth.* It followed that *Nasa poissoniana* must be moving that quickly for a reason. "We thought: maybe they can control it," Henning said. "Maybe they recognize how often pollinators come."

In 2019 Henning and Weigend's newest finding added a stunning layer to the flower's elaborate sexual accounting. After that first bee leaves with all of the nectar, the next bee to arrive won't get any. But the *Nasa poissoniana* will have raised a new stamen full of fresh pollen anyway, dusting the bee regardless. It's a long-established fact that an insect that finds no nectar will not try another flower on the same plant. Instead it will fly farther away to a neighboring plant, carrying the pollen from the empty flower with it and fertilizing the flower on the next plant. That fake-out is the key to *Nasa poissoniana*'s genetic diversity. But Weigend and Henning noticed that the stamen would already be raised before the next bee even arrived. It seemed to happen shortly before the bee's arrival, as if the plant could predict the future. But in reality, it was simply recording the past.

The pair set up an experiment to test if this could possibly be true, playing the role of the bees themselves. In one group of flowers, they probed their nectar cavity every fifteen minutes. In the second group, they probed them every forty-five minutes. A third group was left alone as a control. The next day they came back, and watched as the fifteen-minute group energetically raised their stamens on a rapid schedule, while the forty-five-minute group waited longer, raising their stamens further apart. They tested them again, and found that if the interval

---

* Still other quick-moving plants move for reasons scientists still can't discern. The starfruit tree, for example, moves its leaves all day, for unknown reasons. A prayer plant, a common houseplant, is part of a large club of diverse plants that close their leaves at night, though scientists still debate why they do this. The leaves of a fire fern, which is not a fern but rather a member of the oxalis family, appear to slowly "dance." Again, no one knows why.

between pollinator visits changes—say, from forty-five minutes to an hour and a half—by the next day the *Nasa poissoniana* will have adjusted the timing of its display to line up with the new schedule. It was learning from experience.

"They obviously are able to count the time between the visits and keep that memory," Henning says. Botanists had never noticed this behavior before. *Nasa poissoniana*, in addition to being a masterful pollen accountant, was also a memory flower.

We keep walking along the garden paths. I want to know what Henning thinks about the debate erupting in botany circles in recent years about whether plants can be considered to behave, and whether their behavior could connote a form of intelligence or consciousness.

As I'd discovered over and over again, this was the talk of the town, and very touchy territory. Are plants intelligent? And if they are, are they also conscious? I want to know what Henning, who has just discovered that the plant he's studied for twenty years is capable of memory, in particular thinks of this. The Andean flower was counting time and then changing behavior according to the actual scenario it experienced. Henning and Weigend had called it "intelligent" behavior in their paper, but the word was still enshrined in quotations. I thought it was possible that Henning saw the flower's apparent ability to remember as a hallmark of consciousness—or he could just as easily see the flower as an unconscious robot with a preprogrammed suite of responses. We call our robots "intelligent" too, sometimes.

Memory has long been tangled up with how we think of our own consciousness. Our "sense of pastness," as it is sometimes called, fills out the awareness of ourselves as beings who move through time. Memories are the backbone of the narratives we tell ourselves about ourselves; nothing could be more essential to conscious experience. But philosophers of mind tend to distinguish that sort of long-term memory from the kind that botanists have thus far found plants are capable of. Presumably they would argue that a plant accounting for the changing pressure in its growing body parts or the arrival times of bees is not participating in conscious memory. Yet this is hardly a settled opinion; plenty of other philosophers argue the opposite, that all memory shares a common basis with consciousness. All memory transforms a neutral world into a playground of personal meaning. Of course, this debate is likely to continue as long as the neural mechanisms underlying consciousness continue to elude scientists.

Henning shakes off my question the first two times I ask. But the third time, something in him shifts. He stops walking and turns to answer. Either he is fed up with me, or I've worn out the facade of careful reserve expected of professional researchers. The dissenting papers, he says, are all focused on the lack of brains—and no brains, they wrote, meant no intelligence. "Plants don't have these structures, obviously. But look at what they do. I mean, they take information from the outside world. They process. They make decisions. And they perform. They take everything into account, and they transform it into a reaction. And this to me is the basic definition of intelligence. I mean, that's not just automatism. There might be some automatic things, like growing toward light. But this is not the case here. It's not automatic."

Henning returns to my first question, about where *Nasa poissoniana*'s memories could possibly be stored. That is, of course, still a mystery. But, Henning says, "maybe we are just not able to see these structures. Maybe they are so spread all over the body of the plant that there isn't a single structure. Maybe that's their trick. Maybe it's the whole organism."

Memory, even in humans, is still mostly shrouded in mystery. Neurobiologists have found ways to "see" certain human memories on brain scans, as particular connections of neurons, but many more are yet invisible to science. And then there are the memories the human body keeps but which have nothing to do with neurons at all. Our immune cells remember pathogens, and draw on those memories to respond the next time they appear. Epigenetic memories in cells can be passed down through human generations; we now know that the toll of stress and trauma—as well as exposure to things like air pollution—travels down the bloodline to children and grandchildren, potentially impacting things like inflammation markers. The body, it has been said, keeps the score. But these aren't the type of memories we include in the landscape of our consciousness. The memories our body keeps for us are only silent until they emerge as a change in our health. Then they become very tangible. But epigenetics itself is a field we are only beginning to peel back the curtain on. We don't yet have the words to assimilate it into our sense of ourselves. Memory, even in us, is a tricky thing to parse.

Plants have this cellular sort of memory too. Shortly after my visit to Berlin, I got to experience it myself. I was living on my friend's farm, having emptied out my Brooklyn apartment and moved to the rye fields I'd run through as a child. By now I'd quit my job and committed to reporting out my own questions about plant science full time. Lincoln, the farmer's son, had been in the grade beneath me in school. His parents had watched us both grow up. It was his farm now. It sat on three hundred unruly Connecticut acres of maple forest and pasture an hour and a half outside New York City. Lincoln had two goats, a dozen laying hens, and a gigantic male turkey who forever changed my understanding of his species. He was a magisterial dinosaur, a baroque, charismatic creature whose moods were clear to me. How had I ever eaten such a magnificent beast? Halfway through my stay, this formidable turkey was himself eaten by a bobcat.

These were the cold months. I arrived in November and stayed through January. In the first week of December, Lincoln, his partner, his father, my partner, and I planted the hardneck garlic. Garlic can go into the ground in October, or November, if you're procrastinating, but much longer and you're pushing your luck; the ground still needs a little heat in it to convince a clove it's safe to put out roots. This year it had been put off until the last possible minute. The morning after planting we woke up to the first snow of the season. It blanketed everything. Our two furrows were like a double row of stitching at the hem of a white sheet.

The night before we planted, we all sat on stools in the kitchen, a bushel of garlic between us, separating each head into cloves with the flat side of a butter knife. We looked like oyster shuckers. Freed from their layers of paper with a pop of the knife, the cloves looked like pearls, all smooth milky curves. It is very often the case that I am taken aback by the forms nature makes. How did this garlic make its wispy paper? And these cloves, so perfectly smooth and segmented like an orange made of blond wood, as if each piece was turned on a lathe? Most remarkable of all, though, was that each clove, if nestled point-up in the dirt before true winter set in, would multiply itself. It would sprout white noodle-like roots and green tender shoots, and by July, if all went well (and it usually does with garlic), there would be a whole head of garlic where that single clove was planted.

What the garlic needs, in order to sprout, is the memory of winter. That the spring eventually comes is not enough to make life emerge—a good long cold is crucial. This memory of winter is called "vernalization." Apples and peach trees won't flower or fruit without it. Tulips, crocuses, daffodils, and hyacinth, often the first blooms of spring, need a good strong vernalization too. If you live in a warm climate and buy tulip bulbs, the garden supply store clerk might wisely advise you to put your bulbs in the refrigerator for a few weeks before planting, or you'll never see a flower.

As I sat through the winter, the cold in my bones, I thought about the garlic cloves in the frozen ground, waiting, taking note of the deep freeze that gripped the earth like a vise, counting its passage. Perhaps most instructive of all is that plants know how to wait, how to endure the inhospitable, knowing their time has not yet come but will, and that their flourishing is not a question of whether but when. Thinking of the garlic was a comfort. Their patience bolstered mine. The waiting implied a thing to wait for; the hard ground would thaw, the air would feel like a home for my body again.

The remarkable thing about vernalization is that it means plants remember. The term undisputedly applies; plants use information stored about the past to make decisions for the future. This isn't a singular example. Plants take note of the length of a day and the position of the sun. Cornish mallow, a pink-flowered plant, will turn its leaves hours before sunrise to face the horizon in exactly the direction it expects the sun to rise. The movement itself originates in the tissue at the base of its stalk, where the mallow will adjust the pressure of the water flowing through it to bend in the desired direction. Throughout the day, the amount and direction of sunlight the mallow experiences is encoded in the photoreceptors laid out across its leaves. It stores the information overnight, during which time it will use it to predict where and when the sun will rise the next day.

Researchers have messed with the mallow by simulating a more chaotic "sun," switching the direction of its light source. The mallow learns the new location. This response is "extraordinarily complex—yet extremely elegant," in the words of a research team eager to learn from the mallow to make smarter solar panels.

When it came to my garlic, memory and counting were bound up together. Plants that rely on vernalization must have some way to record the passage of time to be sure the elapsed period of cold—and warmth—is enough. Emerging in a two-day warm spell in February could be disastrous. So they seem to count the days. That's why many plants wait until the warmth is sustained for four days or more. It's less likely to be a fluke.

The fact that plants can remember brings them closer to us, makes them more legible somehow. But it's humbling to remember that they are a kingdom of life entirely their own, the product of riotous evolutionary innovation that took a turn away from our own branch of life when we were both barely motile, single-celled creatures afloat in the prehistoric ocean. We couldn't be more biologically different. And yet their patterns and rhythms have certain resonances with our own. Plants have internal circadian clocks like we do; they need the cycle of day and night. They slow down in the winter and speed up in the spring. They move through youth and old age. And they maintain a record of what they've been through. Memory clearly has deep roots in biology. This makes sense; if the trajectory of all evolution is toward survival, then the ability to remember has a natural evolutionary advantage. It's incredibly useful for staying alive.

As I looked farther afield for other plants that did something like

Henning's flower, memory and movement appeared to go hand in hand. The plant's body records some piece of information, and then moves accordingly. Take the Venus flytrap, the darling of the rapid movers. As mentioned, Venus flytraps can count to five, and can store the memory of that counting at least as long as it takes to figure out if it has a fly in its maw or not. How it works is this: if two of the trigger hairs inside the flytrap's trap are touched within twenty seconds of each other—a good indication that a living creature is moving around inside it—the trap snaps shut. But the flytrap keeps counting after it closes. If the trigger hairs are disturbed five times in quick succession—eliminating all doubt that it has caught a living, wriggling creature—the plant injects digestive juices into the trap, and the meat meal commences. Digestion takes many days, so it's important to be sure.

But if the trap is triggered twice, snaps shut, and the triggering stops, the trap will open again within one day. Clearly whatever is inside is too small to bother with, or not a living creature at all, but perhaps a bit of twig or stone—or, in the case of all the flytraps that have told us anything about their kind, the cold tip of a botanist's probe. The flytrap corrects for its error.

Similarly, some climbing vines are known to count, and also to correct their own errors in judgment, all of which demands memory. Vines have a stronger prerogative to move than the plants that support themselves. Baby vines must seek support structures immediately or risk collapsing under their own growing weight. And so they move, splendidly, flamboyantly, and fast. Once, having neglected a sweet potato on my counter until it had sprouted from several of its eyes, I plunked it into a large pot of dirt and watered it, not knowing what to expect. Within days, a first industrious tendril was grasping a nearby table leg. In two weeks more tendrils joined it. The sweet potato was now coiled up around three table legs and was boldly traversing my butcher block. One tendril had looped itself around a drawer pull. I was delighted to be hosting such an octopus. Why had I only ever eaten a sweet potato and never thought to plant it? This was so much better.

Time-lapse video is a miraculous gift for watching vines and other plants move. Roger Hangarter, a biology professor at Indiana University, maintains "Plants in Motion," an endearing online library of plant movement videos in the style of the early internet. Naturally, I have passed hours there. But the best plant videos on the internet are of the dodder vine, a parasitic plant that takes being a vine to extremes: it grows no leaves, so it must find a host from which to suck out sugars immediately after emerging from the dirt. When it finds a host, it will detach from the ground entirely, relying on its new benefactor for everything it needs. It has no use for chlorophyll, since it does not photosynthesize, so is instead an intriguing shade of orange, and without leaves it gives the impression of a sleek little worm. Watching it grow in time-lapse is a marvel. When the seedling of a parasitic dodder vine emerges, its tip circles the air slowly. It's impossible not to understand that it is looking for something. In fact the gesture looks unmistakably like sniffing. Indeed, the dodder is sampling the air for the emanations of a suitable plant to parasitize. Then, before making physical contact, it begins to move more purposefully in one direction.

The ability to choose wisely is one hallmark of intelligence. The Latin root of "intelligence," *interlegere*, means "to choose between." Dodders are wonderfully fun for watching a plant make a choice: they prefer tomatoes to wheat, for example. Wheat is hard to climb, and not particularly juicy. When a dodder seedling is grown between a wheat and a tomato, it begins to circle the air almost as soon as it pops out of the soil. After a few perambulations it turns, with determination: it has noted its neighbors from afar. Now like a baby snake it crosses the air, aimed directly at the tomato, shunning the wheat. Consuelo De Moraes, an ecologist at ETH Zurich, the Swiss federal institute of technology, was on the team that first noted this phenomenon in 2006. She remembers being shocked at how fast it happened. When she saw them under time-lapse, it reminded her unequivocally of animal behavior.

Once a dodder encounters what it believes to be suitable prey, it begins to wind. Within a few hours, the dodder checks whether it will be worth the effort. Through lab studies, scientists have asked the dodder what it's looking for. The clear answer is the amount of nutritional energy it is likely to be able to extract from that particular host, based on its overall health and the concentration of nutrients circulating through its body. In one experiment, dodders grown among hawthorn plants chose the hawthorns that had been grown with extra nutrient supplements, and rejected the ones that had been grown in a nutrient-scarce environment. Yet it does this before it ever penetrates the plants' flesh, leaving us with questions about how it gathers that information (the likely answer, as with most answers when it comes to plants, is chemical signals). If the prey turns out to be subpar, the dodder stops winding and seeks new prey within a couple of hours. But if it decides the plant is going to make a good host, it begins to wrap more coils around the prey's stem.

The total number of coils the dodder makes reflects the total energy the dodder plans to use to parasitize it. So the dodder counts. More coils means more room for fangs. Once coiling is complete, the dodder sprouts rows of vampiric spikes along its coil edges. The spikes sink into the host's flesh and begin to siphon off its juices. The dodder won't drain enough to kill the plant—a dead host wouldn't be in its interest.* Instead it will keep the plant hobbling along, diminished, but still photosynthesizing. Dodders don't make leaves because they don't need to. They get everything they need from the bodies of others. And they are exceptionally good at it; dodders afflict agricultural fields globally, accounting for severe loss in twenty-five crop species across fifty-five countries. A plant yoked into dodder service just doesn't have what it takes to grow much fruit. Their tendrilic bodies spread out in tangled mats across fields of crops like millions of plant draculae, sinking their tiny teeth into their victim's elongated necks.

The winding dodder, counting its coils, is enacting a sort of living memory through movement. Memory is a way to store what you learn

---

* This is not entirely true, because sometimes the dodder will kill the plant, but then it has done a bad job. It's in the interest of a parasitic plant like dodder to keep its host alive.

about where you live, the dangers and opportunities your world has to offer. Learning is an evolutionary strategy for survival, like the mallow that turns toward the sun. Memory, learning, and movement: they seem bound up together, a package deal.

Anthony Trewavas, the plant physiologist from the University of Edinburgh, is fond of using network theory to think about plants. Though he is certainly trained in the close scrutiny of individual plant systems, he argues for attention to be paid to the whole plant instead—what emerges from the sum of all its parts behaving together. He writes that because animals have always needed to move over a wide terrain to find food, animal evolution "refined sensory and motor equipment and joined the two with a rapid connection," to eventually be joined by nerve cells "compacted into a brain." Can you build a mind from something else? Brains, after all, came into being via evolution, emerging from nonmental ingredients. Flesh, blood, specialized nervous cells.

In 1866 Thomas Henry Huxley famously said, "How is it that anything so remarkable as a state of consciousness comes about as the result of irritating nervous tissue, is just as unaccountable as the appearance of the Djin when Aladdin rubbed his lamp." We've learned a lot about the brain since then. But consciousness so far has proved impenetrable to modern neurobiology. The fact that a brain produces the experience of "mind" is not explained by its mere physical existence.

In Trewavas's assessment, the brain is just one strategy for building intelligence and consciousness. Plants simply took a different evolutionary route, according to their needs: their attention and awareness is localized in each of their parts, but each of their parts communicates and strategizes across the whole, producing consciousness all the same. "The individual plant containing many millions of cells is a self-organizing, complex system with distributed control permitting local environmental exploitation but in the context of the whole plant system," he writes. "Consciousness is thus not localized but is shared throughout the plant in contrast to the more centralized location in the animal brain."

Trewavas points to the way a plant grows as evidence for its constant awareness of itself. Plants are modular. They grow from many nodes, each one tipped in a meristem, a cluster of cells that can be turned into any sort of flesh. The botanist Robin Wall Kimmerer writes that like stem cells, meristems are perpetually embryonic, ready to become what's required. She suggests that these hormone- and nutrient-packed sites of inventive cell-making might be a good first place to look for the seat of plant intelligence. The meristems sense the results of a plant's perpetual full-body scan. The plant monitors each part of its branching body to see how well it's doing—how much each leaf is photosynthesizing, and each root sucking up moisture. If one branch is not bringing in its share, that branch will get fewer resources to keep it alive. Growth, from the meristems, will proceed in other directions. If that node continues to underperform, it will be blocked entirely and allowed to wither, while the plant reallocates its energy to supporting a more productive part of its body.

Once, walking on a road in western Washington in bright midday sun, I came to a stand of gigantic red cedars. Their skirts were so full and thick I couldn't see through them enough to glimpse their trunks. I walked around the corner, stepping inside the stand, and it was like crossing the dotted line between time zones on a map. Night suddenly fell. The dim needle-dusted forest floor had erupted in glossy chestnut-colored mushrooms, floppy as ears. Fallen logs moldered in place, disgorging their red sawdust in damp piles, their disintegrating bark thickly padded in chartreuse moss. The stand of cedars had made this darkness with their bodies. And now I could see their trunks, thick and straight. They had absolutely no branches on this side; it was like being inside a cave made by a waterfall, or an arboreal circus tent. Their glossy green fronds arced out and away from me, toward daylight. Within the stand their trunks were bare, except for a few leafless twigs high up, where perhaps enough light once made it through to bother with a few inward-facing leaves. But you don't move in the direction of darkness, if you're a tree. Those branches were ancient history, shriveled now. I was seeing reallocation in action.

To do this, the plant must remember what is happening in its body at all times, and how long it's been happening for. "The dynamic is continuous throughout the life cycle," Trewavas writes, "requiring a running commentary as it were." The plant runs on a constant stream of choices, finely recalibrating the pressure of the fluids flowing through its own body to account for changes in its form. If a leaf dies, or is taken out of service, so to speak, the pressure in the whole plant system must readjust to maintain equilibrium and remain upright. It's subtle and yet obvious: biological awareness suffuses the whole plant body.

As any plant grows, it builds its body to suit its environment, both in root and shoot. The shape that emerges is a direct response to the physical obstacles it encountered, the distribution of nutrients in the soil, and the direction of the light. In this way a plant's entire body is a physical expression of the conditions, moment to moment, over the whole story of its life. The newest shoots and roots will account for changes; the oldest parts of the plant are a record of conditions that came before. In this way the plant is a map for which one hardly needs a key: the story of a plant's life can be decoded on sight.

In our brains, a memory is a physical entity, a connection between neurons. In plants, memories may be physical pathways too; the root that threads through soil branches and turns, indicating where a patch of moisture once was. The lobe on a trunk where a branch once existed tells the story of sunlight since shaded out. The substrate of these memories is the soil and the atmosphere, rather than our brain matter. The memory is the layout of a physical space, which our own brains are incidentally best at remembering too. Human spatial memory is our keenest form of memory, thought to be a holdover of our hunting and gathering days, when quickly committing the layout of our surroundings to memory was crucial to our survival in a harsh landscape full of danger and reward.*

---

* This is why competitive memory athletes, mimicking the technique of orators of epic poems in ancient times, build "memory palaces" in their minds, placing items to remember as objects in the various rooms of an imaginary house, to be walked through and the objects collected later. For further reading, see Joshua Foer's *Moonwalking with Einstein* (New York: Penguin, 2011).

But what if we hadn't needed to run away from large mammals who were trying to eat us, or if we didn't need to nimbly stalk a smaller mammal across a complex landscape for our dinner? Our brains are centralized and compact, perfect for animals that must travel and take their wits with them. What if instead of hunting, our food was sunlight that rained down on us, so we were bathed in it, and we had to evolve only to be prepared to receive it? Instead of a compact and portable brain, we might instead have evolved a limitless ability to build new arms covered in mouths on short notice. Whereas animals like us live and die by however intact our bodies manage to remain, perhaps we might have evolved flexibility instead of fragility; the ability to nonchalantly drop an arm or two if it is no longer ideally placed to catch food.

If we decide to go down the path of considering the animal brain to be but one format from which to make a mind, and store memories, it might be wise to look for clues in how the mind arose through the evolutionary eons in the first place. In *Metazoa*, philosopher Peter Godfrey-Smith traces the emergence of the animal mind from the first multicellular organisms floating in the sea. When he gets to the first creatures that could swim, or scoot across the sea floor, he wonders if these early animals had awareness of themselves, and their separateness from others. Were they sensing their environment and building memories of it as they went? In other words, were they having experiences? They were certainly moving around a lot, searching for food.

"New and expansive actions carry an expansion of sensitivity along with them," Godfrey-Smith writes. "Nothing is gained biologically from taking in information that is not put to use." If these creatures were moving around, which they were, then they were subjecting themselves to all sorts of new sensory information. It would be a waste not to be able to store that information somehow, to build a picture of the facts of one's world to draw on later. Perhaps the ability to move across space with purpose came first, and the ability to sense that space came after, as a response to the moving. In thinking about what might exist in animals before the evolution of experience, it is interesting to imagine an animal whose motions outrun its senses, he writes. Sensing, and the ability to store those sensations, would tend to automatically catch up. The first animal to experience its world was also likely the first animal who moved by its own effort.

Motion and experience seem naturally coupled. Move, and new experiences open themselves to you. It makes sense to be able to record these inputs and put them to use in our own private quest to thrive. This is the birth of learning. My reading of animal evolution tracks with everything I've learned about plants so far, too. I think about the dodder vine, probing the air for a scent, and then counting out the coils it will need to make its living. As I move about my kitchen, my billowing pothos is keeping track of all its modular parts, making decisions about where to grow next. The answer is always, it seems, toward the window, wherever that might be. And in the mountains of Peru, *Nasa poissoniana*, the memory flower, is raising its stamen—experiencing a bee—and then deciding when to raise its next one to experience another.

Plants, like animals, move through space. But they do it in their own distinct way—through growth. It is easy to imagine them tasting it all as they go, taking account, saving what they learn for later use. The idea of plant memory loses some of its mystical quality when thought of this way.

Memory and experience are intrinsically linked, because a being that can remember the contours of its world can be said to have experienced it. That being is also more likely to make wise decisions—that is, behave intelligently—when faced with the same circumstances later. We learn from our memories. Memory likely propelled our most distant ancestors toward more complex lifestyles that called for yet more complex decision-making. We may not know where plants store their memories—somewhere, it could be said, in their brainless mind—but the knowledge that they evidently have them is enough to change our world. Our private collections of experiences grant us our sense of ourselves, the sense of our own subjectivity that we come to view as consciousness. A more generous view of plant life might extend to them some measure of that same subjectivity; they do, after all, seem to experience and remember their world as they move through it. Mysteries abide, of course. We are far from understanding the extent of memory in plants. We have a few clues and fewer answers, and so many more experiments still to try. But new threads of relation extend between us. A universe of selves comes into focus.

Chapter 7

———

Conversations
with Animals

In *Semiosis*, Sue Burke's 2018 sci-fi novel, plants are central characters. A group of humans have flown through space to escape a war- and climate-change-ravaged Earth to land on a verdant planet, which they name "Pax." They've come to start humanity over, with a different orientation toward the natural world: "With ecology, not against it." They immediately find this means becoming deferential servants to the will of Pax's plants.

"On Earth, plants can count. They can see, they can move, they can produce insecticides when the wrong insect comes in contact with them," says Octavio, the colony's botanist. On Pax, plants have had even longer to evolve. One plant kills several of the colony members when they encroach too far. It sabotages their grain fields. Here, plants seem to be able to strategically plan ahead.

The humans decide that their only chance for survival is to position themselves at the disposal of a particularly powerful vine plant. "We will work for it, not the other way around," Octavio says. "It will help us only because it is helping itself." In a reversal of earthly roles, humans become the "servile mercenaries" of warring plants.

Several human generations later, when another alien species attacks the humans, the vine suggests ways to neutralize the threat without killing them—a Pax precept is peaceful coexistence. "Mutualism can be coerced," the vine says, while contemplating injecting its fruits with stupefacients to disarm the invaders. "In essence, we will conscript a symbiont." The humans respond with stunned silence. The vine adds as encouragement, "This is often done by plants to animals."

BY NOW I'D learned from Rick Karban and his sagebrush about plant-to-plant communication, the elaborate exchange of chemical signals that warn other plants about impending threats. I understood that this quiet speech was everywhere, passing easily beneath our perception, a form of communication requiring neither sound nor movement. A world of meaning was adrift on the air. Yet for the most part, the study of plant-to-plant communication seemed to focus mostly on alarm calls: who warned who, and when. This made sense; preparing for an attack is crucial to survival. But with this much power to transmit meaning, warnings couldn't be the only thing being passed around, I thought. There had to be more to plant communication than alarm calls. This, I learned, was apt—there was much more being discussed. But what I didn't consider, until I came across it, was that plants hardly keep the dialogue to their own kind. They regularly cross the species divide.

Indeed, the relationships plants have with other plant species, and even with animals, is a tapestry of dynamics that run the gamut from reciprocal to exquisitely antagonistic. Often it is hard to tell the difference. The field that studies this is flatly called "biocommunication," but that seems too tidy for the complicated pile of interspecies relationships being uncovered today. As I fall into the world of biocommunication, I get the sense that the rule of life is unruly mixture. Everything, it seems, is impacting and altering everything else. I'm reminded of a phrase used by theorist Donna Haraway, who writes that our lives, whether we notice it or not, are a "rich wallow in multispecies muddles."

Consuelo De Moraes, the ecologist who discovered how dodder vines select their prey, lives squarely inside this muddle. She sits on the edge and watches, squinting, as species upon species interact. In fact, she seems preternaturally attuned to seeing meaningful interactions where scientists before her have seen absolutely nothing. She is a strictly scientific kind of person, and there is a no-nonsense clarity to her speech. She won't tell me about anything she isn't sure she can replicate or doesn't feel confident will make it past peer review. But she also feels capable of wonder, in fact uses it as a primary research tool. Her sense of rigor makes her dispatches from the world of multispecies muddles feel reliable, and all the more gobsmacking for it.

De Moraes speaks to me from across her desk at the department of environmental systems science at ETH Zurich, where she is silhouetted by a gallery wall of butterflies pinned inside square picture frames. These are the adult forms of the caterpillars she studies. She specializes in insects, plants, and viruses that often layer their lives upon one another in ornate ways. For example, in the 1990s, she was studying the triangle of drama between corn, caterpillars, and wasps. First a caterpillar chews a corn plant. The plant notices this, and samples the combination of saliva and regurgitant the caterpillar leaves behind on its leaves; now the plant knows the caterpillar's species—or at least which species of wasp it needs to come parasitize it. The plant then releases a finely tuned chemical gas. Within an hour, the correct wasps arrive. The wasps, no doubt appreciative of the ideal scene before them, insert their needle-like appendages into the caterpillars' bodies, injecting their eggs inside them. When the eggs hatch, the wasp larvae use their especially oversize mandibles to eat the caterpillars from the inside out. They then spin their Tic Tac-shaped cocoons, which they glue to the emptied husks of their caterpillar. The result is a green wormish form bristling with white silky spines, like hedgehog quills made of felt. And thus the plant attempts to save itself. De Moraes discovered this behavior in corn, tobacco, and cotton in 1998.

Two decades later, De Moraes noticed little bite marks on the leaves of some black mustard plants she was growing in a greenhouse. The marks looked like tiny crescent moons, always a tell for a bumblebee mouth. But why were bumblebees biting plants? She kept watching.

The bumblebees, she realized, were starving. They'd been flying around the closed buds of mustard flowers for days, but no bloom had yet opened. If they didn't slip their tongues in a pool of sugary flower water soon, they would begin to slow down, their bodies desperately trying to conserve calories. Eventually they would land on the dirt, crawl a while, and die. The poor bees; their timing was all wrong. The flowers weren't due to open for another month. It wouldn't do. The bees, she saw, began to bite the plants' leaves. The next day the flowers bloomed. The bees drank the nectar and survived.

Interesting, De Moraes thought. Bees don't eat leaves. There seemed to be no good reason to waste precious energy on something that wouldn't feed them. But here they were, biting them regardless. There were moons all over the place. Someone must have noticed this before, she thought. "And then you do this literature search, and it's like, how did they miss this?"

She set up an experiment with all the proper controls and found that bees biting plants made their flowers bloom as much as thirty days earlier than they would otherwise. Obviously, the bees benefit here—but, Consuelo found, so do the plants; they bloom in time to have bees around to pollinate them. Timing in nature is like that: all species rely on other species in one way or another. If they're out of sync, everyone loses. Survival depends on having a way to communicate across the species divide.

"One of my students sent me photos of her spinach salad with these half-moon shapes," De Moraes said. "We would never look for it before, but now we're looking around and going, oh my god, bee damage. After you know, your mind starts seeing that thing all the time."

Common bumblebees, she's found, can tell at a distance whether a yellow monkey flower will have lots of pollen for them or not, by the scent of a certain floral volatile compound that translates, in a bee brain, to "heaps of pollen." But making heaps of pollen takes a lot of resources. So the monkey flower has developed a shortcut. It recognizes the rules of this pre-screening process, and instead of making extra pollen, will exude the volatile anyway—in essence, it will lie. The bumblebee, now duped, will arrive to disappointment. Either way, the monkey flower got what it wanted: a bumblebee to dust with pollen. The monkey flower is an excellent liar.

With biocommunication, it's monkey flowers all the way down. De Moraes talks about the "arms race" of insects, plants, and viruses, each outsmarting the other and being outsmarted at turns. "Everybody is trying to survive. All of them." She often laughs as she explains her findings, as if experiencing her own amazement at the results all over again. I get the sense she isn't as wedded to this vision of nature as a battlefield, but it's a metaphor near at hand in all of science since Darwin, so she uses it.

A few years ago, when her graduate student saw a plant he liked in a greenhouse at a botanical garden in Zurich, he brought a few seeds back to De Moraes's lab. They grew them out and admired the plant's dark purple stems, which were covered in thorns an inch long each. The spikes totally eclipsed its diminutive yellow flowers in size. Every part of it was poisonous to humans. The plants' common names were "purple devil" and "malevolence." They put young caterpillars on them, just to see what would happen. A short while later, they noticed globules of something sticky clinging to the devil plant's stem. They were perfect glowing spheres of sugar, like dew sitting on lady's mantle at dawn. "Extra-floral" nectar, meaning nectar found outside the flower, is not unusual in plants; typically it's there to attract a sugar-eating animal that can help the plant in some way. They kept watching the caterpillars. They were indeed lured by the sugar spheres. But caterpillars don't help plants. They eat them. All of a sudden something was wrong with their mouths. Their tiny hinge-like mandibles were stuck around the globules as though with glue. "It's like if you have taffy in your mouth," De Moraes said. The baby caterpillars waved their heads back and forth, trying to clean out the substance. It was a feeble attempt; without arms that reached their mouths, there was little they could do. The perfect globes of sugar were a trap. Try to swallow one, and never open your mouth again.

Lately, she's been working on goldenrod. Goldenrod, with their tall stalks and gently arching sprays of golden yellow flowers, are nearly ubiquitous in eastern North America. In 2020, De Moraes found that goldenrod can sense the volatile signals of nearby gall-forming flies and jump-start its immune system before the flies ever make contact. Gall-forming insects can be a menace to plants: once they arrive, they hijack the plant's DNA, forcing it to build a house for them out of the plant's own flesh. The architectures that emerge can be ornately geometric, and can come in colors that the plant doesn't normally produce itself. The insects often lay their eggs inside the gall, which can pose a problem for the plant once they hatch hungry. All in all, from the plant's perspective, it's best to avoid such situations. So it makes sense that the goldenrod would have evolved a way to figure out if gall-forming flies were nearby. But the flies, knowing the plants might sense their arrival, also assess the goldenrod. If the goldenrod is exuding volatiles that indicate it has put up anti-fly defenses, the female flies that carry the eggs take notice and avoid it. It's better, for the fly, to move on in search of a less-defended specimen.

Interspecies chatter of this sort is constant and entirely invisible to human perception. The communication is sophisticated, dynamic, multilayered, and quick—all of this happens in a matter of mere moments. Only a sliver of it is so far understood, De Moraes says. "I'm always surprised by how much we don't know."

I THOUGHT BACK to the corn and tomatoes summoning wasps. They were in essence enlisting collaborators. Or, seen in a slightly different way, they were using the wasps as tools. The line between collaboration and coercion is sometimes blurry. The wasps certainly benefited from the relationship too. But either way, it was the plant that proposed the arrangement in the first place. From the plant's perspective, it had found the right tool for the job. I thought about how an ability to use tools is a classic test of animal intelligence. I'd seen videos of crows opening boxes of food with sticks and sea otters using rocks as anvils to crack open mollusks. Was this so different? I kept reading; it turned out it's a story repeated over and over in the plant world.

Bittersweet nightshade, a plant in the same family as tomatoes, potatoes, and tobacco, secretes sugary nectar in order to recruit ants as bodyguards. The ants, hooked on the sticky syrup the plant oozes for them, dutifully pluck off the larvae of the bittersweet's mortal enemy, the flea beetle, which are clinging to the plant's stem. They must be quick, before the wriggling flea beetle babies have a chance to bore themselves into the bittersweet's body and wreak havoc. The ants march the larvae deep into their ant nest. The larvae are never seen again.

Several other plants appear to hire ants in this way, enough for the botanical community to give them their own informal name: ant plants. (And a formal one: myrmecophytes.) Some species of ants can no longer survive without their ant plant, as is the case for the symbiotic ants of the tropical tree genus *Macaranga*, who quickly die out when separated from it. Acacias have a similar relationship; they feed their ants and provide them with specialized places to nest in the hollow thorns on their limbs. In turn, the ants aggressively attack anything that disturbs their ant-plant nest tree. The Wikipedia page for ant plants sports a photograph of three large rust-colored ants on a leaf, circling a smaller red ant. Two of the larger ants each have one of the red ant's front legs in its mandibles, and the third has its abdomen. "Ants collaborating to dismember an intruding ant," the caption offers. The tree has conscripted ants as its bodyguards, and the guards are paid in syrup and lodging.

If plants can't do something for themselves, they find other things that can do it for them. But when those other things are living creatures with their own agendas, that might take a little bribing—or manipulation. Legumes, for example, form associations with bacteria in their roots to lock in a steady supply of nitrogen fertilizer. Their roots are strung with round nodules like lopsided pearls that serve as homes for the bacterial colonies. The bacteria in the nodules fix nitrogen in the soil for the plants, and the plants feed them sugars in return. But this arrangement doesn't always work out as well as the legume might hope. The bacteria are fickle; they don't always do the job, or do it well. Their ability to fix nitrogen varies widely. The legume monitors the symbionts living in each of its nodules, making sure they keep their side of the bargain, demanding a one-to-one exchange. If the plant catches bacteria free-riding, it punishes the delinquent bacteria by choking off the oxygen supply to that specific nodule.

The legume arrangement seems like a straightforward case of a transaction between organisms, with a little mischief and punishment thrown in, but other interspecies relationships can be trickier to parse. Coercion in particular is difficult to determine in a biological system. Who's to say the other party is being coerced and not willingly playing along? What looks like coercion to us can have other connotations for the individuals involved. One case is that of a group of orchids that biologists call "sexually deceptive." These orchids are a brilliant example of how finely tuned plants' grasp of biochemistry can be.

Plants are geniuses at synthesizing chemical compounds. They seem to be able to make whatever chemical mixture is necessary to the task at hand, excreting them as gases through their pores, or sometimes through their roots to infuse the soil. Their precision and aptitude for this is beyond any other organism, and it's fair to say it constitutes a completely additional sense that continues to shock researchers on a regular basis. With each new advancement of gas sampling tools, the picture widens. More subtle and specific chemical inventions come into view. It is a foregone conclusion that plants produce far more complex compounds than our instruments can sense.

But back to the orchids. In Australia, evolutionary biologist Rod Peakall spent more than thirty years studying the ways several groups of orchids convinced wasps to try to have sex with them. The point of this was to coat the wasps in their pollen. Which is to say, in order to have plant sex, the plant pantomimed wasp sex, and the wasp unwittingly had plant sex. The mechanics are a little complicated, but perhaps nothing else we know of shows just how intimately involved plants can get in other species' lives. So we will try our best to imagine it.

Like so many of our strangest plants, a lot of the orchids that specialize in wasp sex are native to Australia. Take spider orchids, for example. Their rather arachnid look comes from having done away with the concept of a normal petal. Instead they grow stringy, leggy strands and append a cocoon-shaped bulb at the end of one of them that bounces heavily in the breeze. The bulb is roughly the size and shape of a specific type of female wasp.

Now picture the wasps. The female of this species cannot fly. To have sex, males fly around, looking for the flightless females who sit, in heat, on plants. The male swoops down, gathers her up in a bear hug, and they fly away like two strapped skydivers to have sex in midair. The orchid's wasp mannequin, then, is designed to fit this system beautifully: the male wasp descends, bear-hugs, and then thrashes wildly, trying to lift the dummy female away. Together wasp and fake wasp bounce on the slender sepal strand until the real wasp knocks into the orchid's middle. Pollen is waiting there to splatter itself on his back. After a while the wasp, perhaps realizing his mistake (or perhaps not), flies off, pollen and all. (Another orchid species with a similar strategy glues a neat yellow package of pollen to the wasp's back. The wasp then looks like it's going to wasp school.) The wasp descends on another supposed female wasp nearby, flails around again, and deposits the pollen in the process. What a selection, I imagine the wasp thinking, buzzing around an orchid patch.

For generations most people believed that the wasps were utterly seduced by the shape of the orchid's pendulum.* But, thought Rod Peakall, wasps have decent eyesight. And really, the orchid that best mimicked the form of a female wasp was lazy with the details. It cast a reasonable silhouette, but up close it couldn't be that convincing. Several other species of orchid with the same pollination strategy seemed to put even less effort into their disguise. No wasp is going to look at

---

* In 1928 pioneering Australian naturalist Edith Coleman did conclude scent must be involved. Yet still the chemistry remained elusive, and the idea of visual deception held strong. Edith Coleman, "Pollination of an Australian Orchid by the Male Ichneumonid *Lissopimpla semipunctata*, Kirkby," *Ecological Entomology* 76, no. 2 (19): 533–39, doi:10.1111/j.1365-2311.1929.tb01419.x.

that and be fully convinced, he thought. Plus wasps only conjugate when the females are in heat. This must be a question of pheromones.

Semiochemicals are any compound that is synthesized in one body and released to infiltrate another. They are by definition any chemical one creature makes and exudes to take the reins of another creature. There is no intention implied in the term *semiochemical*, and no malice, just the haunting fact that whether they want to or not, whoever inhales the semiochemical will be made to behave. They may even think it's their own idea.

In the early 2000s, Peakall set out to assess how much this orchid deception had to do with chemistry. The orchids were exuding a convincing bouquet of wasp smells, he thought. He knew that certain orchids only attracted certain species of wasps, so the chemistry must be rather specific. He figured they had to be using some combination of the more than 1,700 floral scent compounds already known.

"We could not have been more wrong," Peakall said, in a keynote to the global community of botanists at the discipline's annual conference in 2020. Almost all the semiochemicals he and his team analyzed were entirely new to plant science. And these were just a handful of orchids. How many more compounds were out there, at work in the air, manipulating the environment in subtle ways, as of yet invisible to us? Not only that, but thanks to recent advances in gas-sensing technology, Peakall was able to see that the wasp attraction depended on the orchid concocting an exact ratio of two or more of these compounds. In each orchid, the recipe was different. One species of orchid synthesized a gas that was an exact 10:1 ratio of two compounds. Another orchid used a 4:1 ratio of two entirely distinct compounds. All of them were new to science. The specificity was mind-boggling and threatened to exceed even the most advanced modern tools to detect it. Plus, they found, the orchids needed UV light to make their semiochemicals work at all. In other words, the orchids used sunlight as an ingredient.

Peakall started coating little black beads on sticks with these subtle compounds, to see if they would attract the wasps without the dummy. It worked like a charm, proving that chemistry, not visual deception, was the magic trick. "Of course it remains a mystery how in evolutionary terms orchids can intercept and co-opt the private communication signals of their wasp pollinators in such a precise way," Peakall said. It was hard to imagine a more finely tuned coevolutionary system between plants and insects.

But as I listened to him speak, I also wondered what might be missing from this story. It certainly looked like sexual deception, but what if the wasp knew the ruse and was playing along? Natasha Myers, an anthropologist of science at York University, and Carla Hustak, who studies the history of science at the University of Toronto, offer a different way of seeing the relationship between orchid and wasp. The orchid and insect bodies, as even Charles Darwin noticed in 1862, were exactly articulated to one another, the most perfect adaptation in nature. But he still saw it as essentially deceptive; since the insect was not getting any reproductive advantage from the encounter, the orchid must be deceiving it. But, ask Myers and Hustak, could this be something else? Perhaps something other than Darwinian survival-of-the-fittest mechanics? They suggest a sort of flirtation between insect and orchid, an interspecies dance in which both parties consent to the arrangement and move toward it with pleasure. Perhaps these encounters could be proof of a different sort of ecological arrangement, where "pleasure, play, or improvisation within or among species" is the norm. Could the wasp, essentially, be said to "'indulge' in the pleasures of pseudocopulation"?

It might sound like a stretch, but then again, Darwin himself experimented with orchids in his home, engaging in a sort of multisensory experiment, prodding, poking, and brushing orchid parts with his finger, threads of hair, and other tools to see if he could elicit the same response as insects, causing their pollen sacks to eject themselves onto him. He essentially spent long swaths of time role-playing very horny wasps. He described orchids as having their own rather sensual affinities for certain kinds of touch, obvious even to him, a human. Could, ask Myers and Hustak, the relationship be seen as a reciprocal entanglement, in which both the orchid and its pollinator are mutually satisfied? Darwin, after all, spoke of nature as an "entangled bank," an "inextricable web of affinities" between species who were highly involved in each other's lives. There might be room here for another, less antagonistic, more intimate reading of plants' involvement in the lives of other beings.

It must also be said that research has found that these orchids tend to slightly underperfect their chemical mimicry, subtly changing some aspects of the chemical concoction to be convincing but not absolutely indistinguishable from the real thing. This makes sense: if they outperformed the female wasps, luring males too well, perhaps male wasps would be so entranced as to not ever copulate with a real female wasp at all. The orchids would risk losing their pollinator. One wonders what the wasp, who might notice the discrepancy, thinks of all this when he decides to copulate with a flower anyway.

In *Braiding Sweetgrass*, Robin Wall Kimmerer, the botanist and member of the Citizen Potawatomi Nation, describes wanting badly as a young woman to know why asters and goldenrod tended to bloom together each September. The blazing yellow goldenrod beside the royal violet asters produced an intoxicating visual dynamic. Why, she wanted to know, were they so beautiful? She was a Native kid who spent her childhood using personal pronouns for nonhuman creatures. When she entered university in the 1970s, her advisor seemed to take a special interest in rustling that out of her. She was told that her question about beauty had no place in a botany department. Beauty was subjective, not objective, and therefore could never be a suitable line of inquiry for science. Plants must be objects, not subjects, to coax scientific answers from them.

But, Kimmerer learned, it was not an unsuitable question for Indigenous science. Later, after gaining her PhD and a faculty position in a botany department, she attended a gathering of Native elders where a Navajo woman spoke for hours about plants and their preferred relationships—who they liked to grow by, why they were so beautiful. She didn't mention asters and goldenrod, but "her words were like smelling salts," Kimmerer said, awakening her from the scientific absolutism she had, by that time, spent so many years living inside. She returned to her earlier question. When scientists want to measure whether a living thing benefits from one arrangement or another, they tend to measure reproduction. So if asters and goldenrod did in fact benefit from growing together, if there were a purpose to all that beauty, they'd probably have higher rates of reproduction when growing together than alone. Because aster and goldenrod reproduction both rely on pollination, she decided to look at bees.

Yellow and purple are diametrically opposed on the color wheel, and produce a reciprocal visual effect: our eyes respond more strongly when yellow and purple are placed together than they would to either color alone. Perhaps, she thought, the colors have the same effect on bees. All flowers are basically billboards to attract pollinators. The glitzier the advertisement, the more bees show up. Bees have a vastly larger visual spectrum than our own and are able to see colors we cannot. Many flowers are painted in stripes, like landing strips or bull's-eye-like targets that only bees can see. But it turns out, when it comes to asters and goldenrod, that what a bee sees and what we see are basically the same. They too see a dazzling display of the combined effect of purple and yellow. When Kimmerer tested her hypothesis—that asters and goldenrod must grow together for a bee-related reason—she found that they attracted more pollinators while growing together than they did alone. More bees visited the combined plots than the plots of only goldenrods or only asters. The visual display did something to entice the bees, just as it enticed her. Their beauty, she concluded, was on purpose. Our own subjective sense of beauty could tell us something true about plant intentions after all.

Beauty is almost always a form of communication. Namely, it communicates "choose me." Aesthetic preference has been demonstrated across the animal kingdom; animals are attracted to what they perceive as beautiful. It's no wonder that plants incorporate beauty for this purpose too. Flowers themselves evolved in order to be beautiful to animals. Originally, most land plants relied on the wind to carry their pollen grains from one plant to the next. But eventually land animals arrived on the scene and began to eat the plants' protein-rich pollen. In the process of their grazing, these animals transported some of the pollen from flower to flower, completing fertilization in a far more efficient and tidy manner than the wind ever could. Soon plants began to morph some of their leaves into colorful little flags—early petals—to better direct animals toward the location of pollen. These petals took on more elaborate colors and shapes, eventually producing symbols visible to the specific anatomy of the eye of their pollinator. Nectar and floral scents added to the display. These enticing configurations emerged as flowers, which pushed themselves to aesthetic extremes in the race to attract creatures with ever more complex eye anatomy and ever more discerning aesthetic tastes.* The beauty of flowers now speaks clearly even to us.

WE KNOW PLANTS' biochemical genius makes them resilient, defends them well against predators, and gets their needs met in subtle and overt ways. And they don't always hew to what a European sensibility might call the "natural order" of things. They don't stick to their own species, or even to some clearly definable gender. After all, the orchid is reproducing through sex with a wasp. Some plants almost exclusively clone themselves, like aspens or dandelions, and still others clone themselves sometimes and have sex other times, like the strawberry. Many plants are bisexual, with male and female genitalia occurring to-

---

* Though, as is true in humans, the initial source of aesthetic preference remains a mystery. Why do we find certain things beautiful? In some cases, beautiful traits may imply some hidden evolutionary advantage, some individual heritable strength. This is known as the "good genes" theory, but it has yet to hold up to study. More often than not, no such correlation between beauty and vigor is to be found. In the majority of cases, beauty is not apparently adaptive at all. One researcher who studies biological beauty called it "truly decadent" in a piece in the *New York Times Magazine*. It's not at all clear why we find certain things beautiful. And yet, being beautiful has clear advantages. It sets off certain things in the animal brain, it activates attraction. Plants, not at all ignorant of this, endeavor to beautify themselves.

gether on the same flower (in plant anatomy these are called, intriguingly, "perfect" flowers). The ancient ginkgo tree can spontaneously switch the sex of a section of its body, producing a female branch on an otherwise male tree. Ginkgo are one of the oldest lineages of trees we live alongside, having persisted for hundreds of millions of years, stubbornly surviving since the time of the dinosaurs. Their sexual fluidity may make them remarkably resilient to whatever the eons have thrown at them.

On a humid June day in Virginia, I walked through the understory of a grove of majestic ginkgo trees with Sir Peter Crane, a paleobotanist and former director of Royal Botanic Gardens, Kew. He was flying to England for Queen Elizabeth's platinum jubilee celebration the next day. But for today, he was among his favorite species. The grove of three hundred saplings had been planted at the Blandy Arboretum in 1929, and the trees were quite tall, though Crane reminded me that he'd met ginkgos far older and taller than these in Asia. The fan-shaped leaves cooled the air inside the grove and tinted the light a pale olive green. I imagined with longing the display these would make in November, when ginkgo leaves turn a bright golden yellow and fall off the branches in a synchronous display, raining down en masse. I'd marveled at this during many a gray November in New York City, where ginkgo are a common sidewalk tree. If you were lucky enough to be in the right place on the right day, it would look like it was raining featherweight gold coins, the sidewalk paved in gold scales. Instead of that kind of ginkgo show, I was in Virginia for the other end of the tree's yearly cycle: hundreds of tiny seedlings were sprouting out of the forest floor, still connected to the pungent seeds from which they sprang, like chicks halfway out of their shell. Almost all of them were doomed; none of the adult ginkgoes looming over them looked particularly ready to keel over and die, which would be the only way a hole in the canopy might open to give any of these seedlings a shot. Feeling sorry for them, I dug up three baby ginkgoes and carried them around for the rest of our stroll, to pot at home later.

Crane, along with colleagues in Japan, was the first to publish a paper on ginkgo sex-switching, after seeing a small article in a local newspaper in Japan about how a locally famous male ginkgo tree, considered a Natural Monument of Japan, had developed a single female branch. When they went to study it, they saw that the branch had indeed begun making seeds. Beyond that tree, only three other sex-switching ginkgoes—one at Kew Gardens in London, one in Kentucky, and one in the very grove where we stood in Virginia—have been described. He explains that the phenomena is for now believed to be extremely rare, but that may be only because hardly anyone has bothered to look. A full-grown ginkgo tree can have hundreds of branches and be very tall, making close observation of sexual traits extremely difficult and expensive. What's more, the window for checking the sex of a ginkgo is extremely short; one has to wait for their sexual parts to make pollen or ovules, and even then finding a single atypical branch on a pollen-filled tree would be an arduous task that's yet to be undertaken by anyone. But Crane, who has written a nearly reverent book on the cultural and biological legacy of ginkgoes, says that whatever they are doing is worthy of our attention. It may be one of the ways ginkgo has survived almost unchanged for hundreds of millions of years.

There is something decidedly queer in all of this—the orchids and aspens and strawberries and antplants and ginkgoes—a sense of sensual entanglement that disregards binaries, runs across the species boundary, and almost gleefully defies heteronormative modes of reproduction. This lens might also help us escape the idea that everything in nature is a battle, with a clear winner. Sometimes it may be an improvisation, or a collaboration, or something else entirely.

WHEN I SPOKE to Jarmo Holopainen, he was about to retire from the University of Eastern Finland, which has for years been a pioneering institution in research on plant interactions. I wonder if there's something particular to Finland that lends itself to innovations in this field—their trees, in particular, are special; the country is home to great stands of aspen, trees that grow in vast clonal colonies, meaning they are in fact all the same individual. Contemplating a single mega-organism opens new space in the mind; if this is all one being, surely it communicates across its many limbs, which in this case are each themselves whole trees. What is the difference between a single mega-organism and a community? I pictured these aspens as Holopainen picked up the phone. His voice was kind and unhurried and consequential. He sounded like a person thinking about the future of a science he had spent his life in. One perhaps imagines, as a young person, that a field might be better understood by the time you retire. Holopainen is instead looking out on a field of unfinished business.

His field is plant volatiles, the chemicals plants use to communicate. In 2012, he and his protégé James Blande published a beautiful paper in which they described plants' biochemical synthesis as a "language," with the various complex combinations of compounds as the plants' "vocabulary." The combinations and proportions of the compounds in the bouquet, they wrote, could be described as "sentences."

"It's like talking, in that sense," Holopainen said to me on our call. In one particularly elegant study, he discovered that silver birch, which grows well in cold northern climates, would sometimes suffer an attack by leaf weevils. He found that these birch trees defend themselves especially well against the weevil when they grow in an area with *Rhododendron tomentosum*, a plant also known as Labrador tea, which has been used as a tea and a medicine by Indigenous groups in the north for millennia. The leaves of birches growing beside them smelled not like a birch at all, but more like the distinctive scent of Labrador tea. Holopainen, working with Blande and their lab mates, found that the fragrance did not in fact originate in the birch. Rather, it was the same substance that makes Labrador tea medicinal. The silver birch absorbed the scent from its plant neighbor; the compounds would stick to their own leaves, acting as a defense when weevils came around. The same compounds protect them both. It's a complete sentence, woven into existence by two entirely different plants.

But lately Holopainen was more interested in the air that carried those scent-sentences aloft. He wanted to know whether all the pollution humans were putting into the air was capable of mucking up those plant volatiles and short-circuiting that communication. The long and short of it is that they are. As the stakes are becoming clear about how crucial plant communication with other species may be, we are coming to understand that we may be muddling their communication in the first place. The pollution steadily filling the air appears to sabotage plants' ability to send and interpret each other's signals. Just as plants' communication can cross the species divide, so can ours. And we're speaking to them in smog.

Plants are adapted to deal with some ozone, he said. But it presents a form of stress, and like all stress, at a certain degree it becomes too much to bear, causing tissue damage at high enough levels. And at the lower, chronic levels plants are commonly exposed to, particularly near urban areas, ozone can stifle the plants' airborne signals, which simply can't travel as far in a haze of pollution.

Second, ozone can change the makeup of the chemical signal in transit, scrambling the message and making it unintelligible. Third, the receiver plant may never hear the signal in the first place, because it might close its stomata—those lip-like plant pores on the underside of leaves that let gases in and out—when it senses ozone, which is, after all, toxic to it. The same seems to be true for carbon dioxide pollution. "We know that when we grow plants in elevated CO_2 emissions the signaling is also diminished," he said. "We also know that they don't keep their stomata open in high CO_2 conditions." Basically, air pollution makes an absolute mess of plant communication. And it's only getting worse.

The cascade effects are daunting. Plant defense mechanisms—like the ability to turn bitter when warned of an oncoming insect attack, for example—are often the main way those pests are kept at manageable levels. This sort of natural pest control doesn't work as well if plants can't pass messages. "Some species of pests that are normally controlled might reach outbreak level," Holopainen said. "It might have serious consequences."

And it's not just plants that stand to lose. As always in the entangled net of life, the lives of more species are at stake. Blande broke it down for me further. Let's say a plant evolves to summon parasitoid wasps to lay their eggs inside its caterpillars, killing the pests. That wasp is likely highly dependent on that plant to show it where to lay its eggs. If they can't find the host plant for their larvae, wasp populations will decline.

Blande found that when black mustard flowers are exposed to ozone, it takes longer for bumblebees, their pollinators, to find them. He has actually followed the bees right out of their hives using Go-Pro cameras to confirm this. Fewer mustard pollinators means fewer successful mustard plants, and, multiplied across the industry, it could mean shortages or crop failure. And there's no reason to believe mustard is the only plant where this situation could arise; the mustard family (which is also the cabbage family) is just a classic favorite of plant researchers to study, as we've seen. There are likely many more interspecies relationships in peril that no one has yet thought to study.

Now Blande is looking at Scots pine, an attractive evergreen tree that grows well all over northern Europe, holding its own even in the Arctic. He chose it in part because no one had yet studied this sort of problem in a conifer. So far he's seen that when weevils start feeding on the stem of a Scots pine seedling, the plant will release volatiles that cause its neighboring Scots seedlings to jump-start their immune systems. It also prompts the other baby trees to increase photosynthesis, possibly as a way to prepare for the incoming attack. After all, photosynthesis is how plants isolate carbon out of the air, and copious carbon is needed to make all of the compounds that plants use in signaling and defense.

But when pollution arrives, everything changes. "It seemed that some of the plants might have managed to respond but a lot just did not," Blande said. The unattacked seedlings don't increase photosynthesis, nor do they start priming their immune systems when there's pollution around. "I would say that it looked very much that there was a breakdown in the interaction."

On top of this, some evidence is trickling in that the way we grow food appears to be getting in the way of plant communication, though this is far from a generalizable situation. Some domesticated plants actually produce more volatiles than their wild varieties. But research has found that commercial varieties of corn are much less able to produce volatile signals than local cultivars when they notice herbivores laying their eggs on them. They are entirely unable to summon beneficial predators. There are, it seems, silent fields of corn, mute in their moment of danger.

This raises questions about whether fields planted with highly engineered plants grown in vast identical plots for food, like the corn, have had communication unknowingly bred out of them. Or perhaps it has been rendered unnecessary by the hand of selection for a plant that is given everything it needs to survive without endlessly defending itself. This is, in part, why so many pesticides are necessary in modern industrial-scale farming: some plants seem no longer so able to do things like warn each other of invading pests, or summon beneficial predators at will.

It's clear that prioritizing plant communication across the species divide has the potential to benefit both plants and people. We are clearly losing the war on pests. The world uses about two million tons of conventional pesticides to control weeds and bugs each year. (The United States alone says it uses one billion pounds a year.) And these aren't onetime applications; most crops need to be doused in pesticide several times a growing season to keep the pests away. As a matter of course, pests evolve to become resistant to pesticides, requiring higher and higher doses, until entirely new formulas must be developed. The consequences of all this to human health can be severe. In the United States alone, as many as 11,000 farmworkers are fatally poisoned by pesticides each year, and another 385 million are severely poisoned but don't die, to say nothing of the birth defects, breathing disorders, and other long-term health impacts of constant exposure to regular doses of the stuff.* Meanwhile rainwater that runs over sprayed fields takes the pesticides off farms and into streams and rivers, which taint water supplies, extending the health impacts to the general population

---

* Incredibly, that means about 44 percent of all farmworkers are poisoned by pesticides every year.

of humans along with fish and aquatic wildlife. There must be another way.

Yet we have much to learn from the crop plants that do retain a good deal of their linguistic capability. We've heard so much about tomatoes' ingenious defense tactics. What can be learned from listening to them, I wonder? Several varieties of beans are also champions at self-defense; work is being done to breed rice plants to include a terpene from lima beans that attracts parasitic wasps. In trials, the altered rice is newly able to summon the predators of its own pests, too.

Some plant scientists argue for exploiting more of plants' natural defense mechanisms to design crops that know how to protect themselves. Some even point to returning to the old knowledge of companion planting, the practice of paying attention to which plants survive and grow better in the company of other plants—their natural companions. Strawberries offer a clear example of the advantages of companion planting. A strawberry flower is self-fertile; it can produce fruit using its own pollen, or essentially by having sex with itself. It also can cross-pollinate with other strawberry plants, though this requires the help of flying insects. Farmers know that strawberries will produce a third more fruit—and much of it higher quality—when planted beside borage, a medicinal herb that blooms in perfect blue stars. The borage attracts the strawberry's pollinator; the better, more copious berries emerge when the sexually adaptive strawberry opts to copulate by way of insects rather than with itself. While home gardeners and Indigenous farmers have used companion planting techniques for ages, it is still a rare approach in conventional, large-scale agriculture.

I think about the goldenrod and asters blooming beautifully together, attracting more of each others' pollinators in the process, and everything that tells us about plant communication—among themselves, between their fellow plants, and between themselves and insects. Plants can ask or demand assistance; the world they belong to appears ready to answer their call. There is a strong argument to be made that we let plants do more of the speaking for themselves.

Chapter 8

The
Scientist
and the
Chameleon Vine

On the plane from New York to Santiago, Chile, the in-seat screen showed me a map of our route as a thick line straight down the globe. We would be in flight for eleven hours, moving directly south. I was reading about the temperate Chilean rain forest in which I was about to spend much of the next week. It was sandwiched between a series of lakes and a chain of volcanoes in a southern region of the long slender country, two hours by another plane yet farther south of Santiago. In 2014 a Peruvian ecologist named Ernesto Gianoli had discovered that a common vine in this rain forest was capable of something no other plant was known to do. It could, quite spontaneously, morph into the shape of almost any plant it grew beside.

Boquila trifoliolata is a simple-looking plant, with bright-green oval leaves in groups of three, like a clover or a common bean. I'd spent hours looking at photographs of it, and felt I knew it well, or rather knew that this oval shape was hardly the whole story. Gianoli had counted twenty different species of plants that boquila could mimic, but the list grew constantly. Whenever he flew down to do fieldwork in this region, he found another. It seemed to be just a matter of looking closely, and time.

Yet despite the fact that boquila had become a minor celebrity in certain botanical circles, Gianoli was still the only researcher who studied it in the place where it naturally grew. He was eager to get back. When Gianoli finally managed to organize a research trip to boquila territory, some eighteen months after I'd first tried to join one of his visits, he sounded relieved and excited. He and his team would be studying another vine while they were there, he said, but I was welcome to join, and there would be plenty of boquila to see.

I'd been following the discovery since I'd quit my job three years prior, and in that time boquila had begun to cause a true botanical stir. A research group in Germany felt sure this incredible mimicry implied the plant could see. How else could it accurately reproduce the texture, the vein pattern, the shape, of a neighboring leaf? Gianoli didn't like that theory. He had a very different idea of what was going on. Something about bacteria, which he would explain to me later. But whatever the mechanism, it was obvious to me that this vine was poised to change our conception of what plants are, and what they can do. It seemed very much worth the trip. I booked a flight immediately.

Gianoli is a professor at the Universidad de La Sirena in Chile, where he specializes in adaptive plasticity, or the ability of plants to adjust their behavior to suit a changing environment. When we began our correspondence, we used voice notes to communicate so that he could respond on his own schedule (his newborn wouldn't sleep, he explained). His voice came through calm, deliberate, with a methodical cadence that gave the immediate impression of a sober thinker. He talked about a childhood spent reading Darwin and playing soccer. He nearly went professional in the sport at seventeen, but chose biology instead. It was an excruciating decision—I felt the emotion in his voice as he described it—but he wanted, like Darwin, to contribute something to the understanding of the world he lived in. His email signature was a quote from Karl Popper, the philosopher of science: "For it was my master who taught me not only how very little I knew but also that any wisdom to which I might ever aspire could consist only in realizing more fully the infinity of my ignorance." After several years of reading about plant discoveries, the infinity of my ignorance—our collective human ignorance—felt increasingly obvious to me.

Gianoli started out studying insect-plant interactions, but found himself more attracted to the plant side of the equation. "Because plants aren't supposed to do 'smart things,'" he said. They were supposed to be passive agents in their interactions with insects. But he quickly saw they were doing "much more than was supposed to happen." Gianoli wanted to know if I'd heard that plants could send out plumes of chemical signals to summon the natural enemies of the insects that chewed on them, to come pluck the insects off and eat them. Or that some plants are able to detect that an insect has placed its eggs somewhere on their body, and sabotage the eggs. "Or now recent findings report that plants are able to listen, in a way, because they detect the sound of caterpillars chewing, and then they change their defense profile," he said. I'd heard all these things, but it was fun, regardless, to be reminded that even botanists found these to be bonkers developments worthy of gawkish wonder. "It's never ending, this process of being amazed by plants," he said.

Perhaps fittingly, he ended up specializing in vines. Vines are forthrightly reminiscent of animals: they climb, often with great alacrity. Gianoli's hero, Darwin, was himself absorbed by the behavior of vines for a time, writing an entire book-length study on the subject in 1865. In *On the Movements and Habits of Climbing Plants*, Darwin watched as dozens of vines went about their business using different bodily techniques. Some vines coiled around objects to hoist themselves up, others secreted sticky adhesive, yet others grew tiny hooks to secure their ascent. And all of them located their scaffolding by way of slowly revolving their growing tips in a circle through the air until they bumped into something solid. It's impossible not to think of orangutans or cats as he watches his plants "scrambling" through thickets of sticks or pulling themselves up over the jungle gyms of twine he built for them. They seemed perfectly able to course-correct, too: when Darwin slipped a stick out of the grasp of a vine's coil, it would simply straighten out the loop it had made and recommence the search for something else to climb.

Darwin used all sorts of species for his study, sourced from the collections at Kew Gardens in England, where naturalists of the time were bringing back exotic species from their journeys by ship to far-flung parts of Asia, Oceania, and Latin America, often in service of English imperialism. In Ceropegia, a vine family from Africa, southern Asia, and Australia with flowers that resemble open parachutes, Darwin watched a growing shoot slowly and gradually slide up a stick, but not manage to pass over its summit. Darwin wrote about this as if watching a determined man climb an unscalable mountain. The shoot "suddenly bounded" from the stick, falling to the opposite side, before resuming its ascent again at the same angle, coiling upward. This cycle of climbing, falling off, and climbing again repeated several times. "This movement of the shoot had a very odd appearance, as if it were disgusted with its failure but was resolved to try again," he wrote.

Another plant, a Mexican flowering vine in the phlox family, grew hooks to climb the scaffolds Darwin built for it. "When a revolving tendril strikes against a stick, the branches quickly bend around and clasp it," he wrote. "The little hooks here play an important part, as they prevent the branches from being dragged away by the rapid revolving movement before they have had time to clasp the stick securely." The hooks remind me of bats climbing craggy walls with their hooked thumbs. The fact that this plant would hold a branch in place in order to keep it from escaping its grasp as it twined around it reminded me of the many parrots I've watched brace a spray of millet with their feet while they use their beak to pluck off each grain. This action of holding something in place is so familiar, so unmistakably creaturely.*

---

* Darwin also happened to watch two plants in the same family as boquila; *Akebia quinata*, a vine with purple edible fruits that taste like chocolate, native to Japan,

So vines have a long track record of incredible feats. But when Gianoli made his own boquila discovery, that this diminutive Chilean plant was a sort of chameleon in vine form, no previous or proven theory in our understanding of plants could explain what this particular vine was doing. Mimicry of this sort has never before been observed in any plant. Figuring out how boquila managed it would, in his words, require departing from "known paths of knowledge." And as with any true unknown in science, there would be tempting pitfalls along the way; explanations that sounded plausible but could derail research for years if they turned out to be wrong. "I think that if we are able to crack this, if we are able to discover the mechanism underlying boquila's capacity to do this, most likely we will get into a new concept. A new process. A new interaction. A new . . . something." He laughed into the microphone of his smartphone, and pressed send.

To appreciate the mystery boquila poses—why spontaneous mimicry defies everything we know about plants so far—we have to back up to the way a plant senses light. To mimic something, you presumably need to know, on some level, what it looks like. Light sensing is the primary way animals know what things look like. We call it vision. Plants sense light too, mainly because they need to eat it, and at times avoid it. But could it also be how boquila is managing its trick?

No other force has a greater bearing on a plant's life than light. But too much light can be dangerous, scorching a plant's leaves. Plants have come up with all sorts of ways to avoid leaf scorch. But light is also the enemy of plant roots, which typically thrive in near-total darkness.

---

and *Stauntonia latifolia*, a Himalayan plant sometimes called the "sausage vine," with plump, oblong fruits that taste like eggplant. (Boquila is native to Chile, but almost all of its relatives are found in Asia.) These were among the fastest twiners Darwin watched, capable of completing a circle in under three hours. It is easy to picture the motions of these vines in time-lapse, but Darwin, predating that technology by a long shot, simply watched for hours and days, marking the vines' progress over time.

In botany labs, plants are often grown in translucent boxes and clear petri dishes, so scientists can watch the roots as they form. Roots are known to grow ten times longer in a laboratory setting than they would in the dimness of soil, or in other words, as they would in the wild. Scientists typically chalk that up to the prime growing conditions of a lab. The thinking goes: good soil, plenty of light and water, why wouldn't a plant grow excellently? But, Slovakian botanist František Baluška, who we know as a member of the early group of self-professed plant neurobiologists, has an alternate theory. The roots are actually just running away. The light is a stress factor, and the roots, which can sense it, are growing as fast as they can away from it. Baluška says this is a major flaw in study design and has potentially sullied decades of scientific literature. He and his colleagues have shown light phobia in corn and arabidopsis roots, and advocate for using darkened petri dishes in lab settings. But he's taken the idea beyond the realm of mere light "sensing." He now suggests we start using different language, something more to the point: the roots can see light. They have, he says, a form of vision.

Two years before my trip to Chile, I met Baluška on the top floor of the Institute of Cellular and Molecular Botany at the University of Bonn, in Germany, where he led a research lab. He was finishing an email, and pointed me toward a seminar room down the hall. It was a cloudy gray day in Bonn, and the sky was leaking in periodic bursts. Only the mosses in the university botanical garden below were enjoying the weather.

Baluška joined me after a minute or two and perched gingerly in a chair, tilted forward in his seat like a runner in starting blocks. He was tall, with broad shoulders and blue eyes. He told me he was retiring the next year after several decades in the lab. He looked at me and asked, What do you want to know? I got the sense that he was bemused by me, a reporter from New York, arriving a bit drenched and trying to dry my damp notebook on the table.

At this point, Baluška was famous among botanists, or infamous, depending on your view, for being a founding member of the Society for Plant Neurobiology, and for experiments where he found it was possible to anesthetize plants. If plants can be knocked unconscious, does that make them conscious? Baluška says absolutely. "I think consciousness is a very basic phenomenon which started with the first cell," he says. And besides, what is consciousness but the ability to handle situations, to take care of yourself? "If you are not conscious, you are not aware of your environment and you cannot act. You are out. If someone is taking care of you, you can survive, but you alone cannot survive." An anesthetized plant is not conscious, he says, and that difference is the whole point.

Then again, who knows? Baluška gestures to the empty space beside him. "You cannot be sure of the consciousness even of your friend. There is no way to prove it. You can just guess," he says. "The only semi-proof is anesthetics. But there is no other way to be sure that the other person is conscious." I picture anesthetizing my friends, just to be sure.

Our conversation turns to crop plants. Right now, Baluška is deep into research on the corn plant, which he calls "wonderful." They might be able to see, at least with their roots, he says. Before we can get too deep into that, though, he asks me if I've heard of Vavilov. I haven't.

In the early 1900s, Soviet agronomist Nikolai Ivanovich Vavilov discovered a strange phenomenon: weeds in crop fields sometimes begin to look like the crop themselves. Original rye plants, he realized, looked nothing like the plump grain that by then was a staple crop in Russia. It was a scraggly, inedible weed. Rye, he realized, had performed an incredible trick of mimicry.

Early wheat farmers, weeding by hand, pulled out and discarded the rye-weed to keep their planted crops healthy. So, to survive, a few rye plants took on a form more similar to wheat. Farmers still extricated the pesky rye, when they could spot it. This selective pressure molded the rye to evolve to trick a farmer's discerning eye. In this case, only the best impressionists survived. Eventually, the rye became so excellent a mimic that it became a crop itself.

"Vavilovian mimicry" is now a basic fact of agriculture.* Oats are a product of the same process; they also got their start mimicking wheat. In rice paddies, the weed known as barnyard grass is indistinguishable from the rice at the seedling stage. Recent genetic analysis found that this weed began to change its architecture to match the rice about a thousand years ago, when rice cultivation in Asia was well underway. In lentil fields, common vetch is a ubiquitous weed that masterfully redesigned its previously spherical seeds to be the same flat, round disc shape as lentils themselves. In that case, the plant needed not to trick a farmer's eye but rather make itself impossible to eliminate from the mechanical threshing process. Winnowing machines simply could not tell the difference between a vetch and a lentil. Weed genomicist Scott McElroy argues that modern herbicide-resistant plants are actually just engaging in Vavilovian mimicry at the biochemical level; they are mimicking the crop plants that have been conveniently engineered to tolerate the herbicide.

Crop science is typically seen as the domestication of scrawny wild species to turn them into plump, useful food machines, a testament to human will and ingenuity. But Baluška objects to it being true "domestication" at all. "Domestication would be when one partner has more influence than the other one. But there is no evidence for this," he says. "A better word would be coevolution. We are changing them, but they are changing us."

Clearly plants are capable of complex manipulation. Baluška winkingly hints at the thousands of natural plant chemicals we unwittingly ingest every time we eat a fruit or vegetable. "We don't know what they are doing with our brain," he says. "We can never be sure, when we are eating something nice and tasty, that there is not something in this tomato or apple that makes us believe it is the best food."

---

* Success would not reach Vavilov soon enough; he was sent to a gulag for opposing the pseudoscientific views of Stalin's pick for agriculture minister and starved to death at age fifty-six.

I think back to the vine that spoke in the book *Semiosis*. To what extent are we, I wonder, yoked into plant service? We know a little now about plants' genius for synthesizing highly complex chemicals in their bodies that can influence other plants and animals in subtle and overt ways. There are presumably thousands of these compounds we might inhale or ingest every day, simply existing as plant eaters on a plant-dominated planet. We know that some plants are hallucinogenic, some are addictive, and gardening is proven to reduce depression. What about the compounds that may be inside an apple, or an ear of corn? The question becomes, in what other ways are they influencing us? A fleet of humans carefully tending a field of crops can certainly start to look like an army of plant symbionts, diligently serving the plants' needs. I think about Vavilovian mimicry: we didn't domesticate oats; oats domesticated us. When I look at a field of cabbage or pumpkin or blueberries, I wonder: Have they conscripted a symbiont, and is that symbiont us?

But of course we both benefit from that particular form of coercion. Perhaps that's the way to think about all these layered entanglements: they can be seen as antagonisms, or they can be opportunities for symbiosis, for mutualism.

"I think the plants are primary organisms, and we are the secondary ones. We are fully dependent on them. Without them, we would not be able to survive," Baluška says. "The opposite situation would not be so drastic for them."

It is against this intellectual backdrop that Baluška comes to the issue of plant vision. Admittedly, the way he speaks is quite different from the sober data-driven language of most research scientists I meet. He's talking more like a philosopher. But I'm intrigued; I think of the long history of scientists who've posed scandalizing hypotheses far outside the mainstream of their time that turn out, eventually, to be true. Perhaps Baluška is one of these. Or perhaps he isn't.

We finally turn to his ideas about sight. Baluška was first tipped off to the subject through his work on corn roots. But, he says, sight is unlikely to stop at their roots. He thinks that the epidermis of the leaves of some plants (the plant's "skin," one might say) may demonstrate a sort of vision, too. And a much more complex vision than just telling light from dark.

Earlier, when I'd read a letter by Baluška and Mancuso in an issue of *Trends in Plant Science*, I nearly fell off my chair. It was titled, "Vision in Plants via Plant-Specific Ocelli?" That innocent question mark did nothing to soften the implications. *Ocelli* is a science term for simple eyes, and Baluška and Mancuso were asking if plants might have them. It included a mention of boquila, which was discovered by Gianoli two years prior to be able to mimic the shape and feel of the leaves of other plants, right down to their color, vein pattern, and texture.

Recent research suggests that an ancient cyanobacteria, an early ancestor of plants, had (and still have) the smallest and oldest example of a camera-like eye. Baluška and Mancuso ventured to suggest that plants, which evolved from the union of that organism with an early alga, might never have dropped that useful evolutionary feature. In their letter, they note that the cells closest to the surface of plant leaves tend not to have chloroplasts—the cell type that enables photosynthesis—despite the fact that, logically, the very surface of the leaf is probably the best spot to use for photosynthesizing. "This phenomenon is not easy to rationalize," they write. Could it be because those cells are used like ocelli? In other words, were they some sort of very simple eyes?

This is not the first time plant scientists have pondered this possibility. But the last person who presented this hypothesis saw it promptly forgotten for a hundred years. Around the turn of the last century, Gottlieb Haberlandt, a fifty-one-year-old Austrian botanist and an accomplished author of several plant physiology books, began to wonder if a plant was able, in some rudimentary way, to see. He published his theory in a new book, *The Light-Sensing Organs of Leaves*, in 1905.

Francis Darwin, Charles's son and a scientist in his own right, praised Haberlandt's book, and referenced it at length, which is how I, a non-German speaker, came to understand Haberlandt's ideas.

"If sense-organs for light exist they must be sought on the leaf-blade," Darwin wrote, paraphrasing Haberlandt's paper. "We should further expect that such organs would be found on the surface." Haberlandt hypothesized a dome-like simple eye, or ocellus, of the sort that Baluška and Mancuso proposed more than a century later. But the idea never entered mainstream botany at the time.

In 2016, when a group of researchers published their groundbreaking paper that found possible camera-like eyes in cyanobacteria, they wrote that their cells act as "spherical microlenses, allowing the cell to see a light source and move towards it."

Knowing that cyanobacteria have the capacity to see opens the possibility that perhaps the plant kingdom, which evolved from cyanobacteria, never actually discarded it. In the world of light and shadow, where all potential friends and enemies use visual cues to hunt and feed and hide, there's evolutionary reason to believe that once an organism has an eyespot, it would hang onto it. After all, human and all other modern eyes likely evolved from ancient eyespots much like the cyanobacteria's.

Of course, evolution does not always tell such a linear story. Plenty of features across all kingdoms of life have emerged and been dropped over many millions of years, only to pop up again, evolving back into being. But although scientists have not yet located ocelli in plant leaves, that doesn't mean they aren't there. As Baluška and Mancuso argue, no one has yet properly looked.

Vision is fundamentally the perception of light and shadow. Objects become visible to us and other animals when they reflect light back to us. Color, too, is a basic trick of light: it emerges when an object absorbs certain wavelengths of light but not others, reflecting those back to our eyes, determining the color we see it as. Green leaves, for example, appear green because they absorb red and blue wavelengths, returning to our eyes only the green. Chlorophyll in the plant eats that red light to convert the CO_2 and water it absorbs into sugary food; this is photosynthesis. Light includes a spectrum of colors, some visible to us and others beyond our visual spectrum; picture the rainbow that a prism, which splits the light waves, can cast on your wall. When light passes through the green flesh of plants, the plant absorbs some of the red light in the spectrum for photosynthesis, so the remaining light that passes through the plant will contain less red light once it gets to the other side. That means the light that has passed through a plant will have a different ratio of colors; specifically, the ratio of red to far-red light wavelengths—a form of red light on the very extreme edge of our visual spectrum—is reduced. In 2020, researchers found that parasitic plants can read this changing light ratio to know who or what is nearby. In a lab, parasitic dodder vine seedlings appeared to detect the size, shape, and distance of neighboring plants, and used that information to decide which plants to grow toward and parasitize. This makes sense; dodders don't photosynthesize. As seedlings, they have very little time to locate a good host before they run out of their built-in supply of energy. And once a parasitic vine commits to winding around a host plant, their fates are forever entwined too. Choosing the right one quickly is an absolute must. Growing blindly in a random direction would be disastrous much of the time.

To the researchers' surprise, the dodder's assessment of red light ratios appeared to be exquisitely fine-grain. In the lab, they used a combination of far-red LED arrays and real plants to set up tests; when given the option of LEDs arranged to resemble light passing through a grass-shaped plant and another resembling the body of a branched plant, the seedlings chose the direction of the "branched" one (dodders can't grow on grasses). They also chose to grow toward the nearer of any two same-sized plants, even if the difference in distance was only four centimeters. It's not a stretch to say that this parasitic plant can, in this basic way, see its host—or at least the size and shape of it.

But plants don't just have receptors for red light. Botanists have thus far found fourteen types of light receptors in plants, each of which contribute vital information; some allow a plant's shoots to grow toward light, and others help it avoid damaging UV rays. But many of the photoreceptors remain unexplained. In one 2014 paper, botanists in Argentina determined that a few of those photoreceptors were involved in the ability of arabidopsis to recognize their kin; they found the wispy plant detected whether the plant body beside it was kin or

not based on the quality of light passing through it. They were sensing their shape—and therefore somehow, the researchers presumed,
their genetic relatedness—with their photoreceptors. The arabidopsis
accordingly adjusted its growth to avoid shading its family members.
And it's not necessarily that arabidopsis is special in doing this; it is just
the model organism on which botanists often try their experiments
first, which is why we've seen this plant in particular in so many different scenarios, across labs and objectives.

By the mid-2010s, the idea that plants could see had begun to bubble up among the plant neurobiology set. The mechanics of how plants
were sensing subtle differences in the visual field were a mystery, as
was how an overall image could be integrated into a response without the sort of centralized processing normally done by a brain. And
then, in a Chilean rain forest, Gianoli made a startling discovery that
changed the landscape yet again.

Gianoli was taking a walk on a break from the relentless field collecting during a research trip with his students when he noticed something
strange. A leafy shrub seemed to be growing out of the ground from
two stems, one much thinner than the other. He took a closer look. The
thinner stem was not the same species as the shrub at all—it was *Boquila trifoliolata*, a common climbing vine in this part of the forest.
But, astoundingly, the boquila's leaves were shaped exactly like those
of the shrub. He had looked at boquila countless times; it was everywhere in this forest. But he'd never noticed this feature.

He soon found another small tree ensconced in boquila, and again,
the boquila leaves were shaped like that tree's leaves. After a moment,
the gravity of what he was seeing hit him. He knew this was something
big. "It's hard to put it in words. It was a kind of an emotion. I realized
that this was a discovery," he said. "What is the dream of a kid that likes
science? To make a discovery, right? A dinosaur bone or whatever. And
it was close to that. Close to that dream of the kid. But for it to be really
fulfilled, I need to see the mechanism elucidated."

Once he knew what to look for, boquila was everywhere, and everywhere taking a different form. It was astonishing. "By then I knew

the basics of mimicry, all the tricks species do," he said. In every case, they're the result of generations of slow change. "Because of that I understood at that very moment that this was extraordinary, because it was a within-generation response. It was not the result of generations and generations of a sustained effect, but rather a plastic response."

No one had studied boquila on its own; it only grows in Chile and had not been, until then, seen as particularly remarkable. He walked back to the hut where the rest of his team was waiting and turned to his undergraduate student Fernando Carrasco-Urra. "Quieres ser famoso? Do you want to be famous? I've got the idea for your thesis."

After this trip Gianoli and Fernando published a suite of remarkable boquila findings; a single vine plant could climb and imitate up to four different trees' leaves, including their shape, color, texture, and vein pattern. Sometimes, if the leaf in question was particularly complicated (say, a sawtooth serration pattern along its edge), Gianoli and Carrasco-Urra would find the boquila had "done its best," producing a half-serrated, lopsided-looking leaf, like an amateur sculptor trying to imitate a Michelangelo. This trick seemed to serve the purpose of reducing how much herbivores were able to eat the vine; by blending in with the much more abundant leaves of a tree, each boquila leaf had a lesser chance of being nibbled. But the mechanism—*how* boquila pulled this off—was still a total mystery. Boquila, a truly dynamic chameleon, was the first species found to imitate more than one other plant.

There is only one other plant known to do anything close to this. Mistletoe, our sometime symbol of romantic love, is a parasitic plant. Like all parasitic plants, it sinks tendrils into its host plant, often a eucalyptus or acacia tree, and sucks out the nutrients it needs to survive, instead of making them itself. Very romantic.

But some mistletoes take this parasitism a step further by morphing to look identical to the host plant. Not only do they appropriate the host's hard work, they appropriate its identity, too. In photos of mistletoe growing on eucalyptus in Australia, it is virtually impossible to distinguish between the two plants. The mistletoe in this case grows

leaves that are the exact same stiff, round, silvery wafers as the eucalyptus. In another photograph of a sprig of she-oak mistletoe beside an
Australian river she-oak, both plants have long, drooping needle-like
leaves that resemble the pin feathers of a parrot. The mimicry is total.

In the case of mistletoe, one species is mimetically paired with only
one species of host; the Australian she-oak mistletoe morphs only into
the shape of an Australian she-oak leaf, for example. Mistletoes also
have the advantage of being completely hooked into the circulatory
system of their host plants; a mistletoe sinks its body into the flesh of
the plant it parasitizes, gaining access no doubt to crucial genetic information that likely helps it morph into its hosts' shape. These are very
intimate, specific relationships that evolved over evolutionary time.*

But while impressive, mistletoe's mimicry doesn't hold a candle to
what Gianoli saw. Boquila is doing something else altogether; it can
evidently adapt to whatever sort of plant the environment throws at it,
and doesn't seem to require physical contact at all. It is sensing neighboring plants in real time and morphing its body to match them—
sometimes transforming its leaves to match the foliage of multiple
different trees at once, without touching any of them. This of course
makes the vision hypothesis tempting.

At the time, Gianoli speculated that the boquila was somehow

------

* Other types of tight evolutionary relationships have produced other examples
of bogglingly specific plant mimicry: common lungwort in southwest England will
dapple its leaves with dozens of white splotches that look uncannily like bird drop-
pings. These might protect the plant; animals may be less likely to eat the leaves
if they appear covered in a vector for disease. Several species of passionflower in
South and Central America will grow leaves that look as if they are adorned with
small yellow orbs. These closely resemble butterfly eggs; a butterfly is less likely to
deposit her eggs on a leaf where eggs have already been laid, to avoid putting her
baby caterpillars through unnecessary competition when they hatch. If a butterfly
passes over the leaf, believing it unsuitable for her clutch, the passionflower has just
spared itself the tribulation of being a first meal for dozens of hungry caterpillars.
See Edward E. Farmer, *Leaf Defence* (Oxford: Oxford University Press, 2014), and
Lawrence E. Gilbert, "The Coevolution of a Butterfly and a Vine," *Scientific American* 247, no. 2 (1982): 110–21.

picking up the information about leaf form through airborne cues, or perhaps some sort of horizontal gene transfer. The plants weren't connected through their roots, so root-to-root communication was out of the question. But when Baluška and Mancuso took a look at Gianoli's research a few years later, it seemed obvious to them that the boquila was collecting the information by sight.

Gianoli himself protested Baluška and Mancuso's assertion. Either horizontal gene transfer or communication by airborne cues were more likely causes, he wrote in a rebuttal. But neither quite made sense, at least not at first glance. Gianoli's own work had shown that when the host tree was totally devoid of leaves, the vine leaves adopted their own normal, oval-shaped leaf morphology; plus, the vine always mimicked the leaves closest to its own body, whether or not they were actually part of the climbed tree—in cases where an overhanging branch of another tree brought its leaves nearer to boquila, the vine imitated those instead. "Vision seems to us a more parsimonious explanation of this complex phenomenon," Baluška and Mancuso wrote in the journal *Cell Press*.

Later, in his office, Baluška told me about Jacob White, a plant enthusiast in Utah he was in touch with who had begun growing a boquila vine on a plastic tree, just to completely rule out the possibility of gene transfer or chemical communication. "He is sending me photos that it is also mimicking this artificial plant," Baluška said, but he'd need to repeat the experiment several times before it could be considered evidence of anything.

I left my visit with Baluška wondering if our definition of sight might at present blind us to the role it plays in plant life. At the very least, most leafy plants already display the bare minimum definition of vision: they are phototropic, meaning they orient themselves toward sunlight. And if plant vision turns out to be more advanced than that, how will that change our relationship to them? I think of cuttlefish, which are colorblind, yet also able to "see" with their skin, instantly mimicking the color and texture of a pile of marine rock or cluster of coral to disappear in the background, hidden from predators in plain sight. As I walk through an empty park later that day, I wonder if I'm being watched.

I STEPPED OFF my long flight to Santiago, Chile, and boarded the second for the last leg of the trip. When I finally arrived in Puerto Montt, it was a damp and cold April day, the tail end of the Chilean summer. Immediately I was greeted by Gianoli and his team: Gisela Stotz, Cristian Salgado-Luarte, and Víctor Escobedo. The group is amiable, happy to be reunited. It has been a while since their last field expedition. They have been working together under Gianoli for nearly fifteen years, each of them studying a different aspect of plant plasticity, the capacity each plant has to change aspects of its body and behavior that go beyond their preprogrammed responses as new conditions arise. They were here to study *Hydrangea serratifolia*, a different and remarkably prolific vine in this particular forest, to see if it was capable of making good decisions about where to grow. But we would see lots of boquila too, they assured me.

We piled into a rental car, and Gianoli drove yet another two hours south through potato fields and pasture, listening to the operatic soundtrack to the film *The Double Life of Véronique*. The Polish director Krzysztof Kieślowski is his favorite.

We arrived at a colony of rustic cabins set around a lake. The innkeepers started a fire in the wood stove and came by later with a hearty dinner, huge portions of potatoes and meat, which the crew washed down with copious red wine while they caught up about their lives, pets and kids and spouses. I slept heavily that night and awoke the next morning to a gray misty day. We packed our lunches and drove into Puyehue National Park. After a short while of walking, we stepped off the trail and into the forest itself. It's rare to have the chance to do this; it is of course not permitted or advised, as a tourist to any forest, to disembark from marked routes. But with this team, who had been coming to this forest for a combined lifetime worth of years, it felt like a rare honor, and likely benign. Still, I had brought a whistle; my other experiences off-trail in forests with scientists had taught me how easily one could wander off and get lost. Even now, I noticed how if I let the team get around a corner ahead of me, I found myself suddenly, entirely on my own. The rain forest overwhelms the senses. Calling out is futile unless you're in very close range; the rain and the birds absorb the impact of your voice, and the dense overlap of green flora absorbs the visual field entirely.

I caught back up with the group, who were taking a water break. Escobedo snapped off squares of chocolate and passed them around. The forest floor was covered in brown marbles. Salgado-Luarte picked up a few, cracked one open with his thumbs, and popped the white meat of it into his mouth. He handed a second nut to Stotz, who showed me how to do the same, and told me they were called *avellanos chilenos*, or Chilean hazelnuts, though they weren't hazelnuts at all, but rather the nuts of the gevuina tree. I was somewhat wary of eating something from the forest floor, as the only serious disease one can catch in this forest is the hantavirus, which is transmitted via rodent urine, but seeing the group enjoying the nuts was too tempting. What the hell, I thought. I popped one open. The perfect white nut inside was creamy and sweeter than a hazelnut, and made a crisp crack when I bit down on it, a satisfying contrast to the sogginess of the vegetation all around us.

Every so often we walked through a tall thicket of bamboo so dense and straight that I felt as though I were a tiny mite walking through a toothbrush. I assumed this bamboo was invasive, as bamboo often is in the United States, and at first paid it little mind. But Escobedo explained that this was quila, and native to this forest. Nothing here wasn't native, he said. No invasive species had yet gotten a roothold in this place, a rare and precious thing in the modern age. Escobedo showed me how to pluck the growing tip of a young quila shoot, the most tender part of the plant, not yet woody. It slid out of its joint with a pop, and Escobedo stuck the soft end of it in his mouth, indicating that I should eat the very tip. I picked a shoot of my own. It tasted pleasantly sweet and green, and reminded me of the ends of honeysuckle blossoms that I nibbled off the bush as a child. For the rest of the day, whenever we passed through a quila thicket or under an avellano tree, I couldn't resist gleaning a shoot or plucking a nut off the forest floor, enamored with the bounty of these wild foods.

Later I saw Escobedo crush a dark-green serrated leaf from a small tree and carry it with him for a while, smelling it. The next time I saw the same plant, I tried it too; it smelled musky and intense but light and clean all at once, like a good Italian amaro, or like mint and or-ange peels warmed over a fire and then dragged through cold clean dirt. Escobedo told me this was called tepa, and said he wanted to turn it into a perfume. "Yeah, he said that six years ago and hasn't done it, of course," Salgado-Luarte teased. The group joked like this all day, rag-ging on each other for their faults as only people who've spent lots of time together in labs and cabins over the years can. "We call ourselves the 'caravana de fracaso,' " Gianoli explained later. "The disaster cara-van."

We stopped at a large tree. It looked suitable for the group's proj-ect, so they got to work. They were studying *Hydrangea serratifolia*, a remarkably prolific vine. Their pink runners threaded the first cen-timeter of soil everywhere I looked on the forest floor. They slithered straight through clumps of moss and over fallen sticks, crisscrossing each other, pointing—heading—in seemingly every direction. These wormlike protrusions are the juveniles of the plant. The adults were draped around the trees surrounding me, thick, woolly, wizened vines with hard skins that acquire an ecosystem of their own. Moss and li-chen hung on them like clothes.

The group was trying to make order out of chaos, to prove that these baby vines are actually proceeding across the forest floor with inten-tion, seeking suitable trees to climb by sensing the shade they cast to judge the tree's size. When the ground-slithering young vines finally reach a tree, they change from pink to green and switch strategy from shade-seeking to sun-seeking. Guided by sunlight, the vine climbs hundreds of feet up the tree until it breaks through the canopy, bursts into crowns of white flowers, and rains seeds back down to the forest floor, beginning the cycle again. If a tree doesn't cast enough shade, it probably isn't big enough to support a vine that can grow for hundreds of years, bulking up to nearly the size of an old-growth tree itself. Perhaps, Gianoli's group thought, the plant accounts for that. After all, everything depends on finding a good tree.

They began planting little flags in the ground next to each infant vine within a two-meter radius of a large, suitable host tree. Their idea was to measure the angles of each baby vine to the tree: if they were on track to reach it, that would be a hit; if not, a miss. If more than half of the vines are hits, that would suggest that their hypothesis might be right; the vines might be actively seeking trees. During a lull in the setup, Salgado-Luarte pointed out a small shrub called *Luma apiculata*. He briskly slapped it and stuck his face in its leaves, so I did the same. It smelled like Meyer lemons and fresh white soap, like the truer version of what laundry detergent tries to be, clean and delicious. "This is how we know it's luma and not another species that looks a lot like it," Salgado-Luarte said, then slapped a neighboring, nearly identical plant. It smelled only green. I thought of the distress signal these smells represented from the plants' perspective, and internally apologized for our small act of violence.

A while later we emerged at a clearing, and I had my first real encounter with what I'd been waiting for. The area was surrounded by a green wash of temperate thicket reaching taller than my head. I stepped closer to the clearing's edge and looked down, hoping to adjust my eyes to pluck out a few individual plants by contrast with the ground. Delicate tendrils of *Boquila trifoliolata* came into view. It crept along the forest floor at the base of trees, looking much like itself, with the straightforward simplicity of its three-lobed leaves I'd seen so many times in photographs. I was delighted to finally see them in the flesh. I followed a few strands of their stems with my eyes as they twined themselves up the bushy tangles of other plants, all reaching well above my height. As they climbed, the boquila's leaves slipped several times from my view. They tucked themselves discreetly between the leaves of the plant they clambered up, and when I went to look for them, pulling back a leaf here and there, I saw that sure enough, they'd taken on different shapes. Boquila was everywhere, and everywhere mimicking whatever plant it had taken for a neighbor. Nothing could have prepared me for seeing a plant morph itself into the near-exact replica of another. I'd been talking to researchers about it for the better part of two years, but seeing it in person filled me with awe that such a thing is even possible.

As we moved along the thicket, I saw that boquila didn't always mimic anything at all. Sometimes it was just itself. But over and over, Gianoli pointed to clumps of boquila mimicking different species. Each time it took my eyes a moment to pluck out the boquila from the plants surrounding it. The replicas were close, but not perfect. Sometimes the stem was the wrong color, or the thickened leaf had not quite thickened enough to pass. On one plant, the boquila leaves were suddenly huge and fingerlike, dark-green and glossy, elongated to nearly the length of my hand. They were matching notro (*Embothrium coccineum*), a species of small evergreen tree that hung its branches into the clearing beside it. Less than five feet away, the boquila's leaves were suddenly petite and wispy, no longer slender fingers but instead round and about the size of a quarter, perhaps fifteen or sixteen times smaller than those of the boquila nearby. Instead of glossy and dark, these leaves were matte and a shade of cool mint green, like a different plant nearby. It was hard to see where the boquila vine began in the dense tangle of green, but Gianoli said he wouldn't be surprised if both types of leaves were growing from different parts of the same plant. In between their two transformations, the boquila leaves took on more of their standard shape, cascading in crisp ovals of kelly green.

We walked a little farther. I saw that where a plant was yellowing, the mimicking boquila yellowed too. Gianoli pointed out a bush covered in shingles of glossy, thick little leaves, dark green, ranging in size from thumbnail to pinkie nail. This, he told me, was *Rhaphithamnus spinosus*. Tendrils of boquila wound around its stem. The boquila leaves looked quite standard at the base, but as I looked higher on the vine, where it began to wind through the leafy parts of the rhaphithamnus, the boquila leaves got drastically smaller and took on a dark gloss. In the older branches, the boquila leaves closest to rhaphithamnus leaves were matched perfectly to the raphithamnus in size, color, and shape. But what Gianoli was most excited to show me was how the boquila had developed a sharp thorn at the extreme end of each leaf tip. I hadn't even noticed the pointy tips on the raphithamnus until Gianoli told me to drag my finger along the underside of a leaf. Each sharp end was curled slightly under the leaf like a claw. Boquila, when mimicking raphithamnus, faithfully reproduces this spike, and similarly curls it under. I dragged my finger across the underside of several boquila leaves, feeling for the toothy appendage.

This is remarkable to Gianoli. Whether or not a plant has a spiny leaf tip, he says, is often used to distinguish the species itself. It's considered central to a plant's identity, an immutable thing that makes it unique. For it to pop up in a plant that has no history of making spikes like this is unprecedented. It would be like a person growing a rhino tusk. It just doesn't happen.

Gianoli also sees the spiny leaf tip as a strong strike against the vision hypothesis—it is impossible to see the spike if you are merely looking at a rhaphithamnus leaf from above. Since the spike is only visible from the underside, how could a boquila growing above the raphithamnus know about the spike, if it was truly using vision to guide its mimicry? At first I agree with him; it seems reasonable. I saw for myself that boquila leaves positioned over raphithamnus leaves still produce a spike. Perhaps it is a hole in Baluška's case. But then I imagine the plant covered in eyelike organs, which is one hypothesis the plant-vision camp put forward. If the "eyes" were everywhere, and the information integrated, then some part of the boquila would be well positioned to notice the spikes, I think.

The boquila, then, disappears into its host. It goes to great lengths to make itself invisible. Why become hard to find? The reason seems obvious: in a world where animals want to eat you, you are less likely to become a meal if you blend into a sea of other identical snacks. Yet maybe that misses another benefit of the arrangement. The boquila, in mimicking other plants, is trying on different evolutionary strategies for life. Each of the plants in the forest have responded to the same environment with a multitude of different designs. Each of them is a physical portrait of a strategy for success, fine-tuned over millions of years. It's a brilliant evolutionary advantage to be able to access the genius of evolution expressed in the bodies of other plants. The boquila is treating the other plants like a living patent library, where all the patents are free to use—at least for boquila.

This sort of interspecies mimicry calls into question the belief that different species are fundamentally different. Yes, in some ways they are. But what if one can functionally become the other with a few tweaks? Categories begin to falter. Lines between species become less absolute. Taxonomy might start to look more like the invention of categories than the discovery of them. An organism that can slip-slide across the species boundary poses a problem to our idea of a fixed form, of predetermined and immutable identity.

AFTER STOPPING AT a waterfall for another chocolate break, the team began to hike up a winding path. I lingered behind, interested in the small plants growing out of the rocky outcropping to our left. Wild fuchsias with their magenta and violet bells sprouted from the rocks. There were several species of ferns, my favorite thing to see. Some were filmy and translucent, only a single cell thick. Others looked more like the sturdy deer ferns I knew from the Pacific Northwest. Then I spotted a series of delicate maidenhair ferns, with their striking ginkgo-like leaflets all parakeet green, hung on shiny black stems. I leaned in closer to inspect them and saw a frond that looked slightly out of place. The leaves looked normal, but the stems were green instead of the maidenhair's black. It was a boquila. I called Gianoli over. He was excited. "That is the first case we know of boquila mimicking a fern," he said, grinning. For a moment I thought he was being a little patronizing, throwing the reporter a bone like that. "No really, that's yours. We'll cite you in the paper." I was delighted. That it was so easy for me, on my first boquila hunt, to contribute a new finding underscored to me just how much more is left to be discovered about this remarkable vine. As I turned out the lights to sleep in my cabin that night, visions of boquila drifted to the fore, each taking a different shape.

The next day, back inside the national park, we stopped in a place on the edge of the forest where the ground had been mowed in sections, leaving behind islands of vegetation. Boquila seemed to thrive in those islands, rambling over plants and mimicking them expertly. The vine seemed most exuberant here, more so than in the denser parts of the forest just a few meters away, where it still appears everywhere but is less showy and full, and, importantly, seems less intent on mimicking its neighbors.

The group stood in a circle in the forest, discussing why that might be. Perhaps the abundance of light in the clearing helps the boquila to see? Stotz joked. You know, "vision," she said, with air quotes. But she had a point: the amount of sunlight that streamed in here, unlike in the forest, might mean the boquila could make more energy and had more resources to do costly things like change leaf shape and color and vein pattern. This is Stotz's wheelhouse. The whole group, and Stotz specifically, works on plant plasticity, meaning a plant's ability to express a bigger range of its possible forms. Plants with more access to resources are known to be more plastic, pulling out all their behavioral stops. A plant with more sunlight or nutrients can be, in a way, a fuller version of itself. Salgado-Luarte, meanwhile, studies how plants growing in the shade tend to hunker down and brace themselves, waiting for better times ahead. He is particularly interested in how leaves of some species in this forest expand extravagantly in the sun, increasing their surface area to drink up as much light as possible. After all, in a rain forest, you never know when the canopy will close up again. But the same species, if it finds itself in a shady place, will make small, tough leaves—minimizing their energy use, hoping to simply endure until better times present themselves. If it can hang on long enough, perhaps a big old tree will finally keel over and open a gap in the canopy, bathing the plant in light again before it's too late. Resource availability certainly determines things like extravagance. You have to have the energy for it. And was not boquila's trick the most extravagant of all?

A while later I asked Gianoli what, if not plant vision, he thought was happening with boquila. When Gianoli thinks, he closes his eyes. "Of course any explanation sounds weird, bizarre, strange," Gianoli said, hedging. "But I still think that the most likely explanation involves microorganisms." Gianoli thought microorganisms—likely bacteria—were jumping from the host plant to the boquila, and that those microorganisms were directing the leaves to change shapes by hijacking and redirecting the genes that control leaf shape. Instead of assuming the plant itself was changing its own shape, Gianoli saw it more like a form of contagion, that something from outside the boquila was acting on it, like a disease acts on a body. And what infects plants? Microbes of all sorts. But if this *was* an infection, it had to be a type of infection capable of doing some pretty invasive biological rearranging at a fundamental level. Leaf shape, color, size, and texture are all outcomes of developmental programs inscribed in the plants' genetics. Something, Gianoli figured, must be altering the expression of the genes. Microbes, it turns out, are the only thing presently known to be capable of altering genetic expression in plants.

In the 1990s, researchers discovered units of genetic material called "small RNA," or sometimes "micro RNA." They originate in microbes like bacteria and viruses, and as of now 2,600 distinct types of micro RNA have been discovered in the human body. It is believed that these bits of foreign genetic material collectively regulate as much as one-third of the genes in our genome. Even more recently, researchers have discovered that micro RNA play a role in plants' lives, too. They are often exchanged between parasitic plants and their hosts, and can act as signaling molecules between plants. The small RNA from one plant is also now known to be able to interfere with the gene expression in other, nearby plants.

This, Gianoli thinks, could be what is going on with boquila. Genetic material from microbes could be controlling the part of a plant's genome responsible for leaf shape, and nearby boquila could simply be picking up on the same interference; being showered, as it were, with foreign microbial genetic material.

"I don't like microbes. They're so difficult to deal with, to measure, to control, to avoid. I feel more comfortable with macroscopic things. But the weight of the evidence around several systems convinced me," Gianoli said as we ambled through some particularly dense vines. If his theory is true, it would imply that the general appearance of all plants is controlled by microbes, and that the microbes' sphere of influence extends beyond the plant itself into a sort of cloud. Boquila, in this vision, would be unique only in that it is receptive to the microbial clouds of other species. At every step, Gianoli's theory would rewrite what science believes about plants in general. It's an earth-shattering claim for botany, from start to finish. But then again, so is Baluška's theory of plant vision. In a way, Gianoli's theory is not totally far-fetched; it is an extension of the world of microbial influence that scientists are in the process of uncovering now.

We sit down on a log covered in pink and orange slime molds. Gianoli talks about the fact that termites were recently discovered to have microbes in their guts that make it possible for them to digest the chemicals in wood. In other words, the most signature behavior of termites—wood-eating—is made possible by entirely different organisms living within them. The termite's gut microbes, in turn, are able to function thanks to yet smaller microbes that live within them. The presence of these animals-inside-animals predates the evolution of the termite itself—some termite ancestor likely acquired them by eating dead plant material where microbes were living. From there, both evolved together into the versions of each that exists today. One species of Australian termite is known to have a protist in its gut that in turn carries around four types of bacteria of its own. So many cascading individuals make a termite possible. "They are independent. They're from different families. It's crazy," Gianoli said. Over and over, the thrust of these new discoveries points in the same direction: a termite is never just a termite. The same is clearly true for every organism. "What you thought was managed or the result of the action of this organism, well, it seems at least half of it was done by some bacterium."

These cascading constituents intrigue Gianoli. The termites are composite organisms, made possible by multiple classes of beings working in concert. Humans, too, are composite organisms, he reminds me: our own microbiomes appear to govern many aspects of our health, and possibly even our psychology. "They're related to digestion, allergies, even certain psychological disorders," he says.

A year before this trip, Gianoli and his colleagues had visited the forest to take boquila samples when they noticed how randomly the mimicry seemed to be distributed. "You don't see the mimicry one hundred percent of the time," he says. He saw the mimicry around 70 percent of the time. "The intensity, the magnitude of the stimulus varies. And that's why I thought: an organism could be behind this pattern of very patchy effects." When they brought the samples back to their lab and ground them up, they found the first glimmer of evidence for that hypothesis. The bacterial community of the boquila leaves closest to the shrubs they mimicked closely resembled the bacterial community of the shrubs themselves. Nonmimicking leaves on the same boquila plant, situated farther from the shrub, had a totally different bacterial community. "They were clearly distinct, in spite of the fact that they belong to the same organism and are separated by barely thirty centimeters," Gianoli said. "I think that's incredible. And that's microbes." This by no means proves his hypothesis outright; a lot more work is needed to parse out what's really going on. "But it strongly suggests that microbes are involved," he said.

To double-check, Gianoli would have to grow plants in a lab, which has so far proven nearly impossible; he's tried a dozen times, but the boquila plants always grow badly and die quickly. Plus, the seeds are nearly impossible to come by; he's only caught a boquila with seeds once. "I know other scientists sometimes think I'm trying to hide seeds when I tell them that," Gianoli said. He and his colleagues had just figured out how to grow boquila tissue cultures, to circumvent both problems, and he hoped to start on the lab experiments soon. He was irritated with the bluster of the scientists in Europe, declaring the vine to be sighted without doing the experiments to be sure. That's not how science is done, he said. You have to show it works first.

THE RAIN WAS plunking heavily on the millions of leaves all around us. The rest of the team were kneeling beside a tree a few yards away, measuring vine angles. Gianoli asked me if I'd ever heard of philosopher-biologist Rupert Sheldrake's concept of a "morphogenetic field." I hadn't. He explained that Sheldrake imagined a hypothetical biological field surrounding every organism, like a cloud of information. "It's a type of field of influence," he said, invisible but potent, like a gravitational or magnetic field. The morphogenetic field, as Sheldrake conceptualized it, directed the development of an organism's physical form. I pictured clouds of information encircling the plants all around us, plumes of biological instructions. Sheldrake's vision includes things that Gianoli calls "mystical." For example, Sheldrake believes that morphogenetic fields could be the basis for telepathy. Gianoli was quick to tell me he is not interested in any of that. But the idea of biological fields of influence? That he can get on board with, as a device for thinking about microorganisms' potential influence on plants. "I'm not sure it exists. But I like the idea, the image," he said.

I remember when I first learned that human beings had microbial clouds hovering in the air around them at all times. I'd been sitting at my desk on the fifth floor of a corporate building in lower Manhattan for five hours when the data scientist James Meadow told me I'd probably shed millions of microbes all over my cubicle that day. "You know the dirty kid from *Peanuts*? Pig-Pen? It turns out we all look like that," Meadow said into the phone. He worked at the time at a company in San Francisco focused on monitoring the health of the indoor microbiome in places like offices and hospitals, and he'd recently published a paper. "We give off a million biological particles from our body every hour as we move around," he continued. "I have a beard; when I scratch it, I'm releasing a little plume into the air. It's just this cloud of particles we're always giving off, that happens to be nearly invisible." I looked down at my keyboard and tried to imagine the microorganisms, my microorganisms, marching off my fingertips like passengers off boat ramps. Then Meadow told me my microbes were probably also wafting over to my neighbor's cube. I put the phone receiver down a moment and peered over the gray modular divider at my colleague, unsuspectingly typing just three feet away from me. He looked okay. But, was I wafting?

The recent torrent of microbiome research has revolutionized our understanding of how we interact with the world, as scientists draw connections between all manner of health issues and the creatures living in our guts and on our skin. Our microbes influence our immune systems, our smells, and our attractiveness to mosquitoes. Emerging research suggests they may play a role in autism, depression, anxiety, and possibly even who we are attracted to.

In other words, our microbes might mediate how we think and feel. Our own cells are likely outnumbered by our microbial tenants. Upon closer examination, our individuality—what makes us ourselves—may look quite a bit more like a contained democracy than an autonomous dictatorship.

But microbiomes also extend, very literally, into the air around us, in a sort of microbial cloud. Heat rises. My body heat, Meadow explained, was propelling my biological particles off me and outward all the time. My breath, also included as part of my microbiome, is warm and does the same. Each word I choose to put out into the world comes with a host of bacteria I didn't. The size of my cloud will have to do, in part, with how hot or cold my body is at the moment, he said. (I tend to run warm, I think. I probably have a big cloud.)

The rest is up to the "viscosity of the air," which is telling of the scale we're working with here. "We can only feel air when it's hitting us," Meadow explained. But for something as miniscule as a microbe, air acts more like water. Any minor movement can keep a microbe afloat in a room indefinitely. "The tiniest bacteria can be picked up and stay in the air for hours," he said.

"Spirit is matter reduced to an extreme thinness: O so thin!" Ralph Waldo Emerson once wrote. I looked over at my unsuspecting neighbor again. I was literally all over the place.

As we learn more about the integration between our health and our microbes, they begin to seem more indistinguishable from what we conceive of as ourselves. We are not our microbiomes, but we certainly aren't ourselves without them. But just as our lives are not static, neither are our microbiomes; they fluctuate when we travel to new cities, take a shower, take antibiotics, or become involved with a new lover. A mutable microbial identity to match the rest of our volatile selves. Less static than a fingerprint, less easily pinned down, but maybe more faithful to the reality of our tumultuous biological situation. We are always ourselves, but what if "ourselves" is a shifting composite, impossible to separate from the roiling microbial masses within and around us?

I thought about Buddhist meditation, in which the goal is to dissolve the self. Of course, one must know what the self is before it can be annihilated. The way the "self" is described in Vipassana, a form of Buddhist meditation, is as a collection of tiny, quivering units. Some call them atoms. At the root, though, is the idea that we are not ourselves—rather, we are only the sum of a bunch of individual flecks that happen to be humming along in the shape of a person. The self is dissolved when that is understood. It's also a potent image, I think, for what microbes and their clouds imply.

Gianoli's hypothesis about boquila unsettled my notion of a genius plant. Could this instead be a genius bacteria? Or a genius combination of organisms, in which the plant is included? After all, the plant appears better off for the mimicry; animals are less likely to come by and eat it. Then again, it's also in the best interest of the plant's bacteria to not get eaten, I imagine. For whose benefit, then, is the mimicry really? It could be seen as an ingenious technique for bacterial survival. It just depends on one's perspective. Or perhaps choosing a single perspective is the mistake here. The plant and its microbes are likely inseparable. They are a composite organism, a tightly fused collaboration. I thought of another famous collaboration, the one in which a photosynthetic bacteria came to live within an algae cell, forming the predecessor of the earliest plant.

In the 1990s, pioneering evolutionary biologist Lynn Margulis first popularized the concept of a "holobiont," which she defined as a composite organism made of many organisms working in concert. It includes the microbiome, but also the macrobiome—the larger beings in which and upon which the microbiomes live. Cells with nuclei include all mitochondria and chloroplasts, fundamental to both animals and plants. Margulis hypothesized that they first came into being when microbes of different abilities teamed up, eventually fusing into one entity. She believed that these sorts of symbioses between different organisms may have been more important to our evolutionary history than the slow, random mutation science believed to be the source of all evolutionary change. Her paper on symbiotic origins was rejected by fifteen journals before being accepted to the *Journal of Theoretical Biology* in 1967. It was proven true a decade later, when modern genetic analysis became available and researchers could see, for the first time, that every mitochondria and chloroplast does indeed contain DNA from multiple organisms. At the cellular level, we are all holobionts.*

Yet Margulis's holobiont concept has proven true far beyond the structure of our cells.† Crucial traits of animals, including how fast they

---

* Ironically, Darwin himself seemed to grasp at this fact, almost a century before genetic symbiosis was discovered. "We cannot fathom the marvelous complexity of an organic being; but on the hypothesis here advanced this complexity is much increased. Each living creature must be looked at as a microcosm—a little universe, formed of a host of self-propagating organisms, inconceivably minute and as numerous as the stars in heaven." Darwin, *The Variation of Animals and Plants under Domestication*, 1868.

† Margulis was famous for believing in the primacy of bacteria. They were on the planet far before any larger life forms arose, and were wildly successful, adapting perfectly to the chemistry of the early planet, and in many ways engineering it to suit their needs. Our bodies, Margulis wrote, preserve the conditions of that early Earth. The chemical compounds within us, and especially our watery interiors, could be seen as a replication of the cozy primordial world the bacteria first evolved in. We are, in a way, perfectly designed bacterial vessels. "We coexist with present-day microbes and harbor remnants of others, symbiotically subsumed within our cells," she and her son, Dorion Sagan, wrote in 1997. "In this way, the microcosm lives on in us and we in it."

grow and how they behave, have been found in recent years to be the result of microbial signals. This makes sense, when one considers that animals evolved within a world already governed for billions of years by microbes. In fact, Margaret McFall-Ngai, a renowned expert on symbiosis, believes that the human immune system, which has long been recognized to have a "memory" of its own, may be a kind of holobiont-management system. "A memory-based immune system may have evolved in vertebrates because of the need to recognize and manage complex communities of beneficial microbes," she wrote in 2007.

We larger creatures can only exchange our genetic material through creating the next generation—in other words, having babies. But bacteria have no such constraints. They can swap genes in real time with neighboring bacteria, regardless of whether they belong to the same species or not. In this way, a bacteria can adopt new traits from neighbors, adding to the suite of things they can do. If the genetic properties of bacteria were applied to larger beings, Margulis wrote, we would live in a science-fiction world where people could grow wings by picking up genes from a bat, or a mushroom could turn green and begin to photosynthesize by picking up genes from a nearby plant. This gives me a clearer way to see how Gianoli's theory could work: instead of imagining a foreign set of bacteria hijacking the boquila's ingrained sense of personal shape, perhaps the bacteria that lives within boquila and determines its developmental expression could simply be picking up errant genetic cues from the bacteria doing the same thing inside other plants. "People and other eukaryotes are like solids frozen in a specific genetic mold," Margulis and Sagan write, "whereas the mobile, interchanging suite of bacterial genes is akin to a liquid or gas." One begins to see the world in bacterial terms—a microcosmic sea of shifting identity and form. Under the surface, our bacterial selves are morphing and changing. We are all in flux. Who is to say where any of us begin and end?

As we walk out of the forest, Gianoli tells me about another strange case of plant mimicry. Chile, he said, is home to a second plant in the same family as the boquila, the Lardizabalaceae family. This species in the genus *Lardizabala* is an extremely rare climbing vine, growing only in subtropical Chile and parts of Peru. A friend of a friend told him that their uncle lived in a rural village where the lardizabala grew. Its dark purple fruits figured into the village's medicinal traditions. Gianoli hasn't been to this village yet, but he says that when something is part of traditional knowledge, it's likely built upon years of experience and observation. The local lore was that when the lardizabala climbs onto different trees, the fruit that it bears will have medicinal properties similar to those of the tree it climbs on. "So if the properties are digestive, or something related to heart or blood pressure, or other medicinal properties, it is there too." This suggested a different kind of mimicry entirely. "That the fruit inherits the properties of the tree—this would be amazing," Gianoli said.

THE MORNING OF our last day in the field, we drove out to another part of the forest. The group quickly found a tree suitable for their *Hydrangea serratifolia* project and began planting their little flags again. So far they'd seen many more hits than misses. The data was preliminary, but it was looking good. Yet another ubiquitous plant salvaged from the manmade trash heap of supposed passivity.

I wandered around while they took their measurements. In a clearing I saw a patch of *Ranunculus repens* sprouting from the ground, with a boquila right next to it. This type of ranunculus was introduced here less than a decade ago and is now a prolific weed. The boquila growing nearby had arranged itself as a perfect replica in size and overall silhouette. Its three leaves were placed at exactly the same angles as the three leaves of the ranunculus. The lacy cut-out pattern of the ranunculus leaves appeared to be too much of a challenge for the boquila, though it certainly looked like it was trying. Its edges were indented with a series of lumpy dimples instead.

Yet even these mistakes are remarkable. When Gianoli discovered that boquila had even attempted to mimic the ranunculus, it blew apart any theories that were brewing about the trick being a matter of long evolutionary coexistence. "That weed is not part of boquila's evo-

lutionary history," Gianoli explained later. No evolutionary relationship is formed in ten years. The mimicry had to be happening in real
time; this was improvised, not rehearsed.

Spontaneity is a startling concept to swallow; it insists on a highly
alert subject. And the evidence for spontaneity kept trickling in: just
a week before our trip, Gianoli had received an email from a person
growing boquila in his house in London. He sent Gianoli pictures of it
mimicking his house plant, a tiny-leaved ground cover species sometimes called creeping wire vine, which is native to New Zealand. Gianoli
showed me the photos. There was no doubt; the boquila was mimicking this totally foreign plant, and doing a great job of it. Of course, the
New Zealand plant had rather simple round leaves, not such a challenge compared with the other forms I'd seen boquila take. But more
intriguing was its provenance in Oceania, extremely far from the Chilean rain forest. Boquila grows natively only in Chile, but it could apparently copy plants that had absolutely no connection to that part of
the world. The mimicry phenomena was intrinsic to the entire species,
then, and would express itself no matter where a plant went. That was
spontaneous as hell.

Of course, vision is also a tempting explanation for this sort of spontaneous mimicry. In animals, fast responses to something at a distance
are usually due to the ability to see. The idea has obvious mass appeal.
In the car on the way back to the cabins, Gianoli gets an email from
his former student, who was just contacted by a Russian group designing a "megaproject" on plant vision, based around the boquila. In
Bonn, Baluška and his colleagues are starting to grow boquila plants
in a greenhouse to test their vision hypothesis. If Baluška's team suc-
ceeds in getting boquila to mimic a plastic plant in a controlled environment, their vision hypothesis would surely become more plausible.
And there's certainly no possibility of jumping microbial information
emanating from a plastic plant.

But for now, the mystery continues. Whichever way it falls seems
likely to bring on a new conception of plants entirely. Something else
must be going on within them—or between them—to make this type of

mimicry even a remote possibility for a plant. For now, that unknown sits like an invisible object at the center of a room, the thing that everyone knows must be there but no one can see, at least not yet. Whatever it is will change something fundamental about what we think we know about how plants work. "To crack the code of boquila immediately will lead us to crack a general code of plants," Gianoli says. "It goes hand by hand. Understanding boquila will imply understanding plants. That's my feeling."

Baluška's theory is more clearly a vision of plant intelligence, which initially appealed to me. I want to believe plants can see, and perhaps they can. It doesn't seem a totally unreasonable possibility; they have all those photoreceptors, after all. But Gianoli's theory is a vision of bacterial organization and influence that suggests a bigger interconnectedness that appeals to me also. It centers the composite nature of plants, their status as holobionts, inextricable from the microcosmos that they bathe in and which bathe in them.

Either way, it seems time to dim the lights on the idea of plants as individual entities with neat borders. Where a plant starts and stops is not clearly understood. It may not even be a useful question. Ignoring the many ways plants and their collaborators interact—and, ultimately, constitute the plant itself—leaves us with a very partial view of reality. Plants are composites of interpenetrating forms of life that resist an either-or classification. Perhaps much like us. "The completely self-contained 'individual' is a myth that needs to be replaced with a more flexible description," write Margulis and Sagan. "Each of us is a sort of loose committee."

Chapter 9

The
Social Life
of
Plants

Once upon a time, certain insects evolved to be social in a very particular way. They evolved to be fully devoted, each of them, to the well-being of the larger group in which they lived. Their identities became wholly subsumed into supporting the collective. These are the colony dwellers. Everyone in the colony has a role, and to fulfill it some forgo even the activity most often seen as a marker of biological success: they never reproduce at all. Instead, these insects spend their lives foraging for food to bring back to their nest-mates, whose role it is to have the babies. This turns the idea of survival of the fittest on its head. In a colony like this, self-interest is subverted for the sake of group interest. It doesn't matter if *you* reproduce; the important thing is that the colony does.

An entomologist in the 1960s named this lifestyle "eusocial" behavior, and first applied it to bees who live in hives, with multiple generations that take cooperative care of their young, and have distinct roles where only some reproduce. *Eusocial* literally means "truly social." It's a highly complex social lifestyle, full of defined rules of relationality and collaboration. It has since been found to apply to lots of insects, not just bees; termites are eusocial, as are ants, ambrosia beetles, and at least one type of aphid. A coral reef-dwelling shrimp can be eusocial, extending the concept to the world of crustaceans. And naked mole rats have the honor of being the breakout stars of eusociality in mammals.

Insects, crustaceans, mammals: eusocial behavior must have evolved separately many times. Clearly it is an evolutionary strategy for success, or it wouldn't have spontaneously reappeared and persisted across distinct branches of life. If I've learned anything, it's that if something works well, biology tends to reproduce it across the spectrum of life. A good idea has a habit of showing up again and again. So now when I learn that a trait has evolved separately multiple times, I have to wonder if it has a vegetal equivalent. Until very recently, eusociality had never been found in a plant, but perhaps we weren't really looking.

Enter the staghorn fern. In 2021 Kevin Burns, a biologist from Victoria University of Wellington in New Zealand, was walking through the tropical dry forest on Australia's Lord Howe Island. Trees there are mostly stunted. Staghorn ferns, which normally grow high up on tree trunks, are here conveniently at eye level. Looking at these dense clumps of fern, he had an idea. What if these were in fact colonies? Staghorn ferns are unique in that they grow in round, hive-shaped agglomerations of many staghorn individuals, some shaped like spongy discs that adhere directly to their resident tree and to each other, and some shaped like long, green, floppy antlers. Those long fronds are covered in a layer of wax, making them extremely good at directing rainwater to their base, where the disc fronds readily soak up the moisture. What if some fronds are the equivalent of sterile worker bees in a hive, Burns thought, dedicating their lives to feeding their reproductive kin? Indeed, he found that disc fronds never reproduced, and only some long fronds did. The rest lived to keep water flowing to the roots of the whole colony. Could plants be eusocial too?

Complex sociality is, I believe, a type of intelligence of its own, a type of collective intelligence. It goes beyond the proclivities of an individual and is oriented toward making good choices as a group. Intelligence, meaning the ability to learn from one's surroundings and make decisions that best support one's life, is created in a context. It arises out of a need, through natural selection. In this case the need—retaining water in the harsh environment of the vertical face of a soil-free tree trunk—is for collaboration. For a relational aptitude. For a readiness to negate oneself for the flourishing of the whole. This is, of course, the basis for the concept of a community: cooperation, here, is the highest priority.

Collective intelligence is the foundation for some of the most complex animal societies. All animals that evolved in groups develop behavior specifically to exist in that group. Fish, ants, bees, monkeys, people—we all coordinate our behavior in different ways. We call this being social. What if this coordination among individuals extended to our neural systems? Social intelligence is a new field of research in animals, but early results are finding that electrical activity in the human brain can synchronize between people during various social interactions, like communicating, learning, or collaborating on a task. Their brain waves, or oscillations between peaks and dips in neural activity, appear to align. Brain wave synchronization has been found in bats and primates too, which suggests it probably happens in lots of other animals. Clearly, this is a useful phenomenon. Research has found that teams of people perform better when their brain waves are synchronized, that copilots' brains tend to synchronize during takeoff and landing when collaboration is crucial, and that people who are cognitively in sync report higher feelings of cooperativeness and affinity for each other. Couples with higher brain synchrony report more satisfaction in their relationships, and brains of coparents appear to sync up in each other's presence. Our brains evolved in a highly social context, and we are only now seeing how deep that sociality can go. Perhaps this social, collective intelligence deserves its own attention. We may be missing a big part of the story of our own existence without it.

Plants, too, evolved in groups. Fields, forests, colonies, clumps;

plants have always been part of complex social arrangements where interacting with one's neighbor is a part of daily life. How well they manage these exchanges often defines the outcome of their lives. Survival and reproduction are always social questions. As such, plants are indisputably social beings. They run the gamut of social temperament, too: some live in highly collaborative collectives, like the staghorn fern, where group success takes precedence over the success of its individual members. Others seem to prefer a more solitary life. Still others seem positively conflict-averse, with a remarkable ability to share. Many readily make enemies of strangers, but put a great deal of emphasis on family bonds. In a world of shifting resources, it's best to know who you can count on, and family is often a good bet.

These are all clearly matters of living well among your peers, all familiar concepts to us as social creatures. Plants, of course, do it in their own plant way. It takes only a slight attunement to these vegetal versions to begin to see the richness of their social lives. Botanists are only beginning to do that now. A world of social possibility is slowly revealing itself.

THE DUNES THAT ring the shores of Lake Michigan come as a surprise. The undulating mountains of sand extend in vast plains, cresting and falling like the waves of an ocean on pause. Travel a few dozen miles inland, and you're in midwestern farm fields. But here, on the bank of a mega-lake, is where Susan Dudley made the discovery that plants know exactly who their siblings are.

Dudley, a plant evolutionary ecologist at McMaster University in Canada, was being interminably bitten by black flies in the summer of 2006 while observing her study subject. The American searocket is a rather lowly beach shrub, but becomes more impressive when one considers that it has had to eke out its living in the most inhospitable conditions. For a beach shrub, growing large and majestic is out of the question. Life on a sand dune isn't easy. The wind is constant, the water is scarce, the animals are hungry. Any plant that lives on a sand dune has worked so hard to be there that their mere existence is impressive.

It was in the late 1990s and early 2000s that evidence began to mount that plants could recognize "self" from "nonself." They knew if a nearby branch or root was theirs or someone else's. Not long after, Dudley wondered if plant individuation might be even more advanced than that. If they could recognize themselves, could they recognize their genetic kin? Dudley wanted to know if plants could tell more about their neighbors than just that they had them. Zoologists know that recognizing relatives gives animals huge advantages, evolutionarily speaking. Lots of animals have been shown to do it. Why, Dudley thought, not test plants?

As an undergraduate, Dudley found she didn't much like the gruesome task of slicing into living animals. Live dissections of invertebrates were standard fare in biology departments. So in graduate school at the University of Chicago she pivoted to botany. "Nobody cares if you chop up plants," she said; "they call that making dinner."

She started out working on her graduate advisor's projects, testing how plants changed their height in response to their neighbors. "Plants see each other by the color of the light," Dudley said. Light changes color when it passes through a plant, and the light that passes through different plants is altered each in a different way, far too subtle for us to notice but clearly distinct enough for plants to notice. Plants take note of the quality of light falling on them, and whether that light has passed through a plant before reaching them, which indicates a taller neighbor. They then grow their stems to certain lengths accordingly— taller when there are lots of neighbors around, and shorter when there are not. It makes perfect adaptive sense. If you're at risk of getting crowded out, you grow taller to keep your patch of sun.

"Phytocrome-mediated stem extension" is the official term for this behavior. Around the same time that Dudley was looking at it, researchers elsewhere were discovering that plants had a similar belowground awareness: they knew which roots were their own, and which belonged to other plants, and would adjust their root growth accordingly. It makes sense not to compete with yourself. A formula for neighborly plant behavior was emerging: "If they knew they had neighbors aboveground, they got taller; if they knew they had neighbors below-ground, they made more roots," Dudley said.

With this in the back of her mind, she started working with American searocket in the lakeshore dunes of Indiana. This was *Sand County Almanac* country, not far from where Aldo Leopold wrote his classic of nature writing. It was beautiful, but there were the flies to deal with, and the sand itself. Beach fieldwork is a struggle. "Working on the beach is surprisingly not fun when you get sand in your equipment."

But then Dudley had a thought. The searocket might be the perfect species in which to study whether a plant acts differently with its family members around. Searockets disperse their seeds in two ways: some are spread far away on the wind or afloat on water, and some attach themselves to the mother plants, to be carried into the soil when their parent inevitably decomposes. "When the mother dies, you have all these seedlings come up." It was easy to find siblings growing together.

Dudley was right. When surrounded by unrelated plants, searocket would grow roots prolifically, aggressively expanding into the sandy soil in an attempt to monopolize nearby nutrients. But when they grew beside their kin, they would politely confine their roots, leaving siblings space to make a living beside them.

Dudley chalks up this discovery to her choice to temporarily set aside the usual question of how something benefited the plant. Instead, she simply observed what they actually did. "My innovation is that I was asking how plants behave," she said. Looking at behavior is a different thing from looking at benefits to the plant. Sometimes it can be hard to know what a benefit would look like to a plant, anyhow. Humans don't always know enough to infer these things. But they can watch and take notes about what's actually happening in front of them.

It was the first time a plant had ever been shown to recognize its kin, much less give it preferential treatment. Dudley was briefly flabbergasted; "We are always surprised when we find what we predicted. Nature is so complicated." Quickly the surprise turned to apprehension. "It was satisfying and also kind of scary. It's a controversial result." In science, controversial results are subjected to the most intense scrutiny.

And other academics are skittish to align themselves with results until they're accepted into the mainstream. Without allies, it's tough to do much at all. She published her results in 2007, but knew it would be a while before anyone believed her.

Around the same time, another of her students was working with populations of impatiens, a common garden flower native to Rhode Island. The plants seemed to be recognizing their kin too, and treating them better than strangers. The preferential treatment appeared in its aboveground behavior. When impatiens grew with strangers, they would frondesce as aggressively as possible, unfurling extravagantly to coopt as much of the sunlit space as they could. When planted beside kin, they would kindly arrange their leaves to avoid shading their siblings.

Recognizing your kin makes great evolutionary sense. First and foremost, it helps avoid inbreeding. But more than that, it's part of natural selection; Darwin's "survival of the fittest" includes survival of the fittest genes, not just the fittest individuals. If individuals survived at the expense of their close relatives, their genetic success would be compromised. It's been a named rule in animal behavioral science since the 1960s: Hamilton's rule states that you will behave preferentially toward your family members, so long as the cost to your own well-being doesn't exceed the benefit to your shared genetic line. That risk will be worthwhile, from a Darwinian standpoint, if the number of close family members you stand to save outweighs the risks to your own life. This also means that a willingness to help family exists on a sliding scale according to your degree of relatedness. Or, as British biologist J. B. S. Haldane was rumored to have declared, "I would lay down my life for two brothers or eight cousins."

Hamilton's rule depends also on an organism's ability to cooperate and behave altruistically toward its related kin. To do that, an organism must know who its kin are. We already know orca whales live in complex family pods where they regularly share meals and communicate using their own family dialect, and female baboons spend their entire lives within yards of their mothers, aunts, and sisters, grooming each other and napping side by side. Even sponge-dwelling shrimp are known to collaborate with their family members to defend their sponge-nests. But extending that to plants, as Dudley's work did, is a sea change. Colleagues wrote response papers accusing her of bad study design. As almost always, radical new ideas in science are met with extra helpings of doubt. Conservatism in science is both a safety measure against false ideas and the thorn in the side of new break-throughs. It can be painful for the scientist subjected to it. "It's weird, sometimes upsetting," to publish controversial work, Dudley says, but she gets it: when she first heard that roots could distinguish self from nonself, she had a similar reaction. But she came around eventually. Here, she knew her study design was sound. And she saw what she saw. She figured she'd just have to wait her critics out.

Within a decade, evidence to support Dudley's work began flowing in. In 2017 a researcher in Argentina found that sunflower farmers could get up to 47 percent more oil yield from their plants if they grew them in rows with kin closely packed next to one another. They grew the flowers at densities unheard of in sunflower farming, but instead of attacking each other underground, as closely grown sunflowers were assumed to always do, they did the opposite: aboveground, the sun-flowers tilted their stalks at alternating angles to avoid shading their kin-neighbors. There was no sign that they were robbing each other of nutrients, either. If they were allowed to grow at odd angles, rather than be forced to stand straight up, each flower received more light, and oil production skyrocketed.

Since Dudley's first paper, people like Rick Karban have found kin recognition in their study subjects—he saw it played a clear role in how sagebrush in California defend themselves against insect attacks, with more closely related individuals warning each other first. An arabi-dopsis plant will rearrange its leaves to avoid shading siblings, too. Re-searchers in Buenos Aires have tracked the movement of a single leaf to find that it will shift position within two days if it senses a sibling's leaf beneath it.

Plants clearly can recognize their kin. But how they do this—through which sensory channels—is an ongoing line of inquiry, in part because the means appear to be various. In some cases, the siblings are detected by chemicals excreted by their roots belowground. In the case of the arabidopsis, the way a plant notices that its sibling is beneath it is by sensing the quality of light reflected back. In other words, the sunlight passes through its own leaf, hits the sibling's leaf beneath it, and bounces back up to hit the underside of its own leaf again. Somehow the information contained in that reflectance includes everything the plant's photoreceptors need to decipher the other plant's genetic relatedness.

It seemed any species that botanists put to the test would display some form of kin recognition, and change their behavior accordingly. "Example by example we're building up a body of literature," Dudley says. She doesn't expect that researchers will find kin recognition in all plants. But it does show up in a great many of the ones tested so far.

The implication of kin recognition is that plants have a social life. They are aware of who they are in the company of, and decide how to behave toward them accordingly. Their suite of social dynamics goes well beyond kin recognition, too; carnivorous plants, for example, were recently discovered to have evolved to hunt in packs. Collaborating on catching insects allows them to lure larger prey.

In 2017, Chui-Hua Kong's research group at China Agricultural University proved the "two brothers or eight cousins" hypothesis by showing that preferential treatment of kin happened on a gradient, according to just how related plants were. The team grew over a dozen different rice lines, all in soil taken from a rice paddy on the southern bank of the Yangtze River. The lines were variations on two closely related cultivars, or selectively bred varieties of a single species: half were indica inbred rice, and half were indica hybrid. Each set of lines were the progeny of five cross-combinations of six parents. In other words, each of the indica-inbred lines shared one parent, making them all half siblings, and the same went for the indica hybrids. This meant everyone was related, but by varying degrees of closeness. The team planted them in a variety of different cultures, or arrangements of various lines, to see what they would do. If culture is the idiosyncratic way that beings in a group conduct their business with each other, these were certainly plant cultures.

While each culture seemed to behave a little differently, it was clear that the most closely related cultivars declined to compete with one another belowground. The researchers saw no discernible difference in their root lengths. But as they experimented with cultures of more distantly related lines, they saw antagonism begin to creep into the belowground relationships: the root measurements "consistently increased" in accordance with how distantly related the neighbors were. Kin recognition was clearly afoot. When the team blocked the flow of chemical signals between the roots with plastic film, all kin recognition stopped. This confirmed the chemical nature of the exchange; plant roots exude chemicals that seep through the soil, alerting other plants to their identity at a distance.

Then the team introduced a third type of rice cultivar: japonica inbred. Compared with the lines already involved in the study, japonica was very distantly related. The difference was immediately clear. The presence of a distantly related cultivar seemed to inflame the rice's sense of private property. Lateral root formation among the various indica lines shot up; they expanded their roots flagrantly in the direction of their new neighbor.

The japonica rice, seeing its neighbor as similarly foreign, did the same. The result was lots more roots, and somewhat less fruit. That is, rice planted with distantly related cultivars was so busy with aggressive root-building belowground that they spent less energy building body parts aboveground. Much like how sunflower yields went up when relatives were planted together, cultures of closely related rice had more energy available to focus on rice making. Ultimately the team found that rice yield went up when rice was planted in mixed cultures of closely related cultivars. A diverse mix of half siblings seemed to work better than a monoculture of identical plants; it's unclear why. But the yield definitively went down in a mixed culture of distantly related rices.

In animals, the act of mating is another elaborate dance of social choices, often related to family obligations. In plants it may be as well. Rubén Torices, a researcher at Universidad Rey Juan Carlos in Spain, specializes in plant sexual strategies. He sees these kinds of interactions as explicitly within the realm of social behaviors. "The life of plants within a neighborhood—it's a social question," he says. "We should use social theory." This opinion creates problems for him—social problems—with his colleagues in the plant sciences. "It's like a taboo," he says, to apply social theory to plants.

But he does it anyway. In 2018, he and his team found that flowers will invest more in advertising to pollinators if they're growing among their kin. This was a perfect intersection of sexual strategy and family ties. Pollinators tend to be attracted to large floral displays; this is called the "magnet effect." A cluster of especially colorful large flowers is like a giant billboard to an insect on the nectar prowl. But it takes a plant a lot of energy to make the pigments and build the petal material this requires, energy they potentially can't use for other things, such as making seeds later in its life cycle. There's a reproductive trade-off; bigger, brighter flowers might attract more pollinators, but may also limit the number of offspring they can ultimately birth from those pollinated ovules. Torices and his team found that when they potted the Spanish clumping herb *Moricandia moricandioides* with genetic kin, the plants tended to team up, investing in big, showy displays of their magenta blooms. But when he planted moricandia in pots with unrelated moricandia, they'd produce fewer flowers. After trying different arrangements of relatedness across 770 seedlings, they found the pots most crowded with relatives reliably made the splashiest flower displays. The findings were important first because they showed that floral displays were linked to social context. But second, they pointed toward the possibility that moricandia willingly gives up a portion of its potential reproductive chances in order to draw pollinators in to the group, so long as its group is family. If not, the plant hedges its bets and errs toward selfishness, that is, more seed-making power. Torices says more work is needed to be sure that these trade-offs are actually significant enough to make this claim, but if they are, he says, this could be evidence of a type of familial altruism.

Susan Dudley, whom Torices refers to as "our leader" in this field, is interested in altruism too, which is a known phenomenon in animals. Just because a species is known to treat their kin preferentially doesn't mean every single individual in that species will; some individuals may tend toward altruism more than others. In 2017 Dudley suggested that crop breeders have been going about their business with a huge blind spot. They have likely been selecting *against* altruistic plants, to their own detriment. A field without altruistic plants is a field at war. Like any population in wartime, they'll be thrifty with their energy, and it certainly won't go toward luxuries like making fruit.

Crops are typically grown in cultivars, or variations on a single species, bred with some specific trait in mind. Within a single cultivar, the plants tend to be genetically similar, though not identical. But individual altruistic tendencies may be more clearly distinguishable among them. To develop cultivars in crop breeding, farmers select for the most "vigorous"-looking individual plants in a field. But these are actually the most competitive individuals. The plants with more altruistic tendencies will be more reserved, in that they will tend not to grow aggressively into their neighbor's sun space. So it seems the history of crop breeding has actually helped to reduce altruism, to its own peril, writes Dudley.

If a farmer were to instead select altruistic plants early in the breeding process, it could steer the crop toward allocating fewer resources into competing for space, therefore presumably putting more energy into reproduction—that is, the development of the fruit the crop is prized for. On the flip side, aggressive plants are useful when the aggression is directed to plants outside the cultivar—non-kin plants, including weeds. Choosing the plants adept at helping their neighbors but fighting off intruders might ultimately result in a highly resilient cultivar. In this way, paying attention to individual plants' social traits—might we say personalities?—could have a real benefit to the way we grow food.

A SEED BLOWING across the ground settles on a loamy spot. It's moist, it's warm, the conditions are good; so naturally it isn't the first to settle here. The seed already has a subtle sense for chemical cues that tell it where it is and who is near. It must; to a plant, spatial awareness is everything. It samples the chemicals dissolved in the soil moisture, noting the flavor of its new neighbors. Some of them, the seed notes, are its siblings; seeds that have fallen off the same mother plant. Others are of a completely different species. This plant is still a mere embryo, but it already has a decision to make.

Seeds take the gamble of their lives when they decide to emerge. They can often wait months or years for the right conditions. Those conditions are not only things like moisture and heat; their neighbors are also variables that can impact a seed's potential survival into plant adulthood. A seed evidently knows that.

In 2017 Akira Yamawo, a plant ecologist in Japan, tested this capacity on Asiatic plantains, a weedy, low-growing species just a few inches tall with a skirt of paper-thin leaves the shape of jackrabbit ears (fruit plantains, in the banana family, are unrelated). He planted plantain seeds first with their siblings, and noticed no particular difference in when they chose to sprout. He planted them again, this time beside seeds of an entirely different species—white clover—and still saw no significant change. But when he planted the seeds with both their sibling seeds and the clover seeds, he noticed a remarkable change. The kin-seeds synchronized their sprouting and accelerated it, emerging sooner than they would have if planted alone. If one of the plantain seeds was already farther along in its emergence, the other plantain seed would speed up its growth to match it. In other words, in the presence of an entirely unrelated species, sibling seeds will rush to grow, and coordinate their germination to come up together. This offers a clear competitive advantage; come up first, in a group, and the clover can't possibly crowd you out.

This lured my mind in new directions. The synchrony suggested that the sibling seeds could sense the developmental stage of neighboring kin and change their rate of development to match. Yamawo called this "embryonic communication." It also means that all the parts that make a full-grown plant body—like its roots and shoots and stems—are not necessary to the work of clocking neighbors at a distance. The mechanisms are all there in the embryo. A seed has everything it needs for complex kin sensing.

With Yamawo's experiment we've entered the ecological zone known as the rhizosphere, the world of soil and the multitudinous organisms that live below its surface, in and among plants' roots. There is still much we don't know about soil and its pulsing community of life. There are as many as one billion microbes in a teaspoon of soil. Fungi weave their networks of hair-fine threads through nearly every square inch of ground. And plant roots, swerving and diving in search of food, interact with it all, and with each other.

It's time to think seriously about roots, lest we forget that half a plant's life is lived in the rhizosphere. Roots can be thought of as a mass of many thousands of mouths, each autonomous in its search for nutrition, though also highly coordinated among themselves. Plants develop extremely complex root systems, each spreading roots of all sizes from thick taproot to vanishingly small root hair to occupy soil areas often far larger than the amount of space used by the aboveground parts of the plant. For example, a scientist who counted the roots of a single winter rye plant found that it had 13,815,672 individual roots, spread out over a soil surface area of about 130 times that calculated for its shoots. What we see of a plant aboveground is often far less than half the story.

The lives of these roots are full of relationships with microbes and fungi, the contours and consequences of which are only beginning to be understood. Fungal threads are hooked into the roots of nearly every plant grown in the wild, and may be crucial to the way plants communicate with one another belowground. The amino acids glutamate and glycine, important neurotransmitters in our brains and spines that are recently understood to be important to plant signaling, actually pass between plants and fungi at the junctures where the two connect.

In *Entangled Life*, the mycologist Merlin Sheldrake, Rupert's son, describes how these associations can also determine key aspects of that plant's identity. In one experiment, researchers took a species of fungi that normally lived inside the roots of a salt-loving coastal grass and transplanted it into a dry-land grass that could not tolerate the sea. The ability to tolerate salt is considered a hallmark trait of a species. But suddenly the dry-land grass did just fine in brine. The sweetness of tomatoes, the aromatic qualities of basil, and the properties of the essential oil in mint have all been shown to change depending on the species of fungi the plant grows with. The concentrations of medicinal compounds in echinacea, the aromatics in patchouli, and the antioxidant power of artichoke heads have all been found to increase with the presence of certain fungal associates too. The list goes on and on. Where a plant ends and a fungi begins becomes hard to parse. In fact it seems hardly a stretch to ask whether a plant is itself without fungi.

Some evidence suggests that plants—which initially came on to the evolutionary scene as amorphous greenish blobs of algae—developed their first leggy forms precisely to house beneficial fungi. "What we call 'plants' are in fact fungi that have evolved to farm algae, and algae that have evolved to farm fungi," Sheldrake argues. By the time the first plant roots appeared, the plants had already been associating with fungi for fifty million years. By some scholars' accounts, roots are literally the product of fungal influence, made to stitch plants and fungi into relation.

What's more, these entanglements can be mutually beneficial to both plant and fungi. Fungi, existing in the dark underground, can't photosynthesize. So they get their supply of life-sustaining carbon from their plant associates, who spend all day making carbon-rich sugars and fats out of sunlight and air. In exchange, fungi supply plants with soil minerals like phosphorus, copper, and zinc that they mine from rock and decomposing material, which plants need but can't always get on their own.

These relationships are symbiotic, but that doesn't mean everyone involved benefits equally. Multiple types of fungi may be entwined with multiple types of plants in a system, each with their own signature way of doing things. In some cases fungi has been found to "charge" a plant more carbon in exchange for the transfer of a smaller amount of phosphorus when the mineral is scarce, and to do the opposite when phosphorus is abundant.* Scientists still do not know how fungi manage these interactions, much less how they coordinate their dealings across vast mycelium mats.†

Yet plants have their own strategies to get the most out of these fungal associations: researchers found that plants can preferentially direct carbon toward fungal strains that are inclined to supply them with greater quantities of phosphorus. This suggests that neither plant nor fungi has the complete upper hand in their dealings with each other; trade-offs and compromise abound, and so the long evolutionary relationship between the two kingdoms continues.

There is much to be learned about relationships between fungi and plants, but what social dynamics occur when roots of different plants encounter each other in the rhizosphere? When one or a few roots discover a patch of nutrients in the soil, other roots will redirect to coalesce there within hours or days. Roots that have grown into a patch that is subsequently depleted can be pruned back, and new roots can sprout as new needs arise. The swarming ability of roots—each autonomous but coordinated with the whole—has prompted some scientists to draw

---

* In fact, across a large mycelial mat connected to many plants, Toby Kiers, a professor at Vrije Universiteit Amsterdam, and her colleagues found that fungi could use a strategy of "buy low, sell high," by moving phosphorus from places of abundance to places of scarcity, where it will fetch a higher price. See Matthew D. Whiteside et al., "Mycorrhizal Fungi Respond to Resource Inequality by Moving Phosphorus from Rich to Poor Patches across Networks," *Current Biology* 29, no. 12 (June 2019): R570–72.

† There is much debate about these "market forces" that may influence trading between plants and fungi. Reciprocity seems not always to take place. The complexity of real-world associations between them make it hard to generalize; some species of plant appear not to engage with a tit-for-tat relationship at all, instead providing their associative fungi with zero carbon. See F. Walder and M. van der Heijden, "Regulation of Resource Exchange in the Arbuscular Mycorrhizal Symbiosis," *Nature Plants* 1, no. 11 (November 2015): 15159.

a comparison to animal colonies, like ant colonies, beehives, or schools of fish, each of which are self-organizing networks of individuals. Each ant exists on its own, seeking its own food, but also lives at all times in service of the greater ant community. If one ant encounters a particularly good food patch, other ants will redirect to join it. The colony is constantly in flux and adapts its behavior as new conditions arise in its environment. This "swarm intelligence" involves coordination across multitudinous individuals, each in possession of their own brain, but so tightly networked as to function as a sort of collective organism, a single entity made of many awarenesses. Roots, in many ways, can be described the same way. Each root tip acts as both a gatherer and a sensor, integrating information about the rhizosphere into the whole root system, causing the architecture of the plant's root network to morph and shift shape, like a murmur of starlings or a school of minnows.

JC Cahill, at the University of Alberta, is best known for his work on the concept that roots actively forage for their food. *Forage*—a word he chose very purposefully—implies intentional, directed behavior. In fact, *behavior* is also a term he prefers to use whenever possible, noting that he is "senior enough" to use it comfortably without worrying about the security of his job or reputation. "Animal behaviorists have really good theory," he said, and botanists ought to start looking to them to help answer questions about how plants live their lives. After all, plants seem to echo a lot of behavioral principles typically seen in animals. For example, in 2019 Cahill coauthored a paper with Yamawo that found that when you stress a plant by damaging its leaf veins, it will make bad foraging decisions. Instead of putting more roots in high-nutrient soil patches, it will equally distribute its roots in both low-nutrient and high-nutrient patches. That's inefficient, and uncharacteristic. After a while, perhaps after healing itself a bit, the plant seems to recover its senses, and makes advantageous decisions about root placement once again.* "That's mimicking human psychology," he says. There's plenty

---

* This study also showed that stress is a whole-plant hazard, and that the signals of

of evidence that people make poorer decisions when stressed in some way—say, when they're hungry or tired.

It's worth noting here that Cahill is married to an animal behaviorist. Colleen Cassady St. Clair is a biology professor who studies cougars, coyotes, and bears. They've coauthored several papers, and it is easy to imagine a sort of cross-pollination of ideas at their dinner table. "The way I see it now is we need to stop seeing plants, people, nonhuman animals as having different evolutionary motivations. We have the same ones," Cahill says. "It's not that I see plants and humans as analogous, they're just outcomes of the same thing. Natural selection doesn't care what taxa you are."

Cahill studies the interface between plant behavior and community ecology; who is where, how many of them there are, and why. So what Cahill is looking at is how plants behave with each other, forming social cultures and influencing the makeup of the community. The way roots forage for food has everything to do with their social environment. Roots in the soil approach each other, retreat from each other, avoid each other, touch each other. Perhaps nowhere has that been more clearly elucidated than in sunflowers. We've already learned that sunflowers are deft at spatial awareness aboveground, rearranging the angle of their stalks to avoid shading their siblings. But they are even more precise about their movements belowground. In 2019, Cahill and researcher Megan Ljubotina found that sunflowers will take note of their social environment to decide where to place their roots. Sunflowers have one central taproot, and many branching lateral roots. When Cahill and Ljubotina grew sunflowers alone, they found that the plants would quickly locate a high-nutrient patch of soil and place most of their roots there. But when more sunflowers were added, a clear structure of social etiquette appeared. When a high-nutrient patch of soil lay exactly halfway between a sunflower and its neighbor, the sunflower

---

the wound travel through the entire plant's body. What happens to a plant aboveground affects its ability to function optimally belowground.

would place its roots elsewhere, often extending farther down in the soil to escape competition. But if the sunflower was even slightly closer to the nutrient patch than its neighbor, the closer of the two would not hesitate to grow lots of roots into that patch.

Patch-sharing did occur, but it was all rather polite, especially if the sunflowers had other additional patches in which to forage. In situations in which two sunflowers shared a single patch of nutrients, but other promising patches were also within reach, both parties would place their roots in the shared patch, but the roots stayed in their own zones. Neither was trying to gain a monopoly. Instead of the long roots the researchers saw when a single sunflower was monopolizing its own private nutrient patch, the patch-sharers grew short ones. The sunflowers weren't exhibiting what one might call greedy behavior, even when they technically could. Coexistence appeared to be a stronger urge for the sunflowers than competition.

In this way, the sunflowers seemed to have a high degree of sensitivity to their social environment. In situations where resources were abundant, they went to great lengths to avoid competing with their fellow sunflowers altogether. When resources are scarce, however, it's a different story. Sunflowers are known allelopathics, meaning that they will secrete chemicals into the soil when resources are low to stop the germination of seedlings of other plants. As such, sunflowers are often good guards against invading weeds in garden patches. But how a sunflower can possibly sense very fine differences in the distances between it and other sunflowers in the soil, and triangulate that awareness with the nutrient patch, is beyond Cahill. "The spatial stuff stumps me," he says. "I don't know how they do that."

In the grasslands in Canada that Cahill has studied for decades, he knows there are species that seem to prefer growing together. They form multispecies "neighborhoods," which he tells me is the technical term. It doesn't seem like mere tolerance; they actively seek each other out. They are, Cahill says, coexisting. Coexistence is a powerful concept; it doesn't quite have a place in Darwinian schema of ruthless competition driving all changes in life. "Ecologists have decided neighbors must be antagonistic," Cahill says. But he just doesn't see it in the data.

For the last twenty years, Cahill and an ever-changing cast of his students have been manipulating a suite of seventeen variables on the same two hundred hectares of grasslands in rural eastern Alberta, Canada. They string up tarps to mimic different conditions of shade, they add and remove fertilizers, they over- and underwater, they remove certain species and add new ones.

Every single shift in these variables appears to cause a shift in the makeup of the neighborhoods. A species that was formerly a minority becomes dominant, a dominant species suddenly becomes rare. No single species wins out for very long, and never long enough to take over or eliminate their neighbors. That's led Cahill to come to a very different conclusion than the usual idea that certain species simply outcompete others to win a patch of land. "If you have any natural variation in a system, that should maintain biodiversity and prevent dominance," says Cahill. "Natural systems are really complex." But since the dawn of theoretical community ecology as a field in the 1960s, ecologists have used simplified models to predict what will happen in ecosystems, based on two or three types of lifestyles a plant might have. In Cahill's estimation, that radically oversimplifies the system to the point of being useless in the real world. It doesn't capture the heaps of variables actually at play. "We don't see evidence that there are three ways of life. There are a gazillion."

The most earth-shattering thing that this long-term experiment has taught him is that competition isn't actually all that important. It's surely one driver of change, but it's only one in a profusion of drivers. Plant cultures are multifactorial, like human cultures. Resources like food and water and light play a role, but not just as an incitement to selfishness. Everything Cahill changes in his field in turn causes the grassland community to change. He has seen that if he takes out one species, those remaining don't necessarily increase the amount of soil they stretch out into or sunlight they expand to monopolize. If the plants were indeed in a constant state of competition, you'd expect that the remaining plants would jump into the newly vacated space and gorge themselves when their competitor is suddenly out of the picture. But they don't.

The makeup of neighborhoods rearrange themselves, sometimes in extremely unpredictable ways, and never following ecological models for what is supposed to happen. "You don't need competition to explain this," he says. "It doesn't mean competition isn't there, but we can explain all of our patterns without ever once talking about competition." Why are we giving competition such importance in community ecology? "Just because people said it fifty years ago doesn't mean it's true," he says.

This is a very different way of looking at evolutionary history. This isn't survival of the fittest in the traditional sense. Or rather, it *is* survival of the fittest, but "fittest" here doesn't mean what we thought it meant—it's not about whoever manages to demolish their neighbors. This is more like survival for a while, until something changes. In a way, it's an opportunity to shift our perspective; while changes are causing a complex drama of decline and abundance at the level of individual plant species, in the end, the thing that survives is the biome, the whole community of life, just in varying states of composition. This makes me think of the "chance variation" that drives Darwinian evolution. Darwin made a clear provision for randomness in his vision of how species evolve: a species will move through various random mutations until something offers individuals an advantage. Then that Mad Libs-style mutation sticks, becoming part of its kind. This is a continual process; random mutations are being tried and dropped from—or kept in—species' lineages all the time. Competition isn't the focus, though it is sometimes the factor that makes a new trait worth keeping. Still, change is constant, random, and irrepressible, and it happens to be the dominant force driving species evolution.

Change, for a species or a field, is never finished. Complexity—the marriage of the extreme idiosyncrasy of species and the constant fluctuations of a zillion variables in their environment—may be the whole point. Very little seems to be fully predictable. Even kin recognition is a slippery concept; yes, plants do seem to be helping their kin quite often. But then, sometimes, they don't. Cahill himself has supervised student experiments where the kin effect seems absent, or a plant will be more antagonistic with their kin than with unrelated plants. Just when a rule emerges, it shows itself to be not so easily pinned down as settled fact. Natural systems are really complex, and our theories aren't. And that's the problem. Complexity itself may be the answer.

I acknowledge to Cahill that I feel overwhelmed trying to understand that. Change itself as the driver of ecosystem change? It's a messy concept, recursive, doubling back on itself. It's hard to imagine in concrete terms. "Mentally it is a struggle to get your head around it," he says. "But at the same time, I think that we hurt ourselves in ecology, not just plants, but in community ecology in general, by relying a lot on super simplified models that were great at the start in the fifties and sixties, when they were put forward to help frame thinking of a new discipline. But people still use them and think that they are likely representative of what's going on in reality." Plants are not always at war, and they don't always respond to difficulty the same way.

Increasingly, colleagues in the plant sciences are showing how complex plants really are. The surprisingly adaptive mechanics of their bodies, their ability to precisely respond to their environment, and their capacity for spontaneous decision-making suggests that the old way of seeing plants as simple and predictable organisms must be tossed out. Ergo, the way of seeing them as simple and predictable members of an ecosystem must too. "I think that complexity matters," Cahill says. "That's not the dogma right now in community ecology, but I think it will be in another decade or so. It is hard to get your head around that, and it's nice to simplify ideas. But you know, nature is not simple."

Chapter 10

——

Inheritance

In the eastern Atlantic forest of Bahia, Brazil, on the sandy, mossy ground beside the off-grid house of a single amateur botanist, grows an inch-tall plant with reddish stems ending in tiny dart-shaped flowers. The flowers are white with bright pink tips, like a fountain pen dipped in ink. The whole plant emerges only during the rainy season, springing up within weeks of the persistent wetness that begins in March and dying back entirely by its end in November. Within a month the little dart-flowers open, get pollinated, and disappear, having done their part. Capsules of fruit appear in their place, holding the seeds of the next generation. The usual course of events. But then something unusual happens: the fruit-tipped stems begin to bend toward the earth, genuflecting, craning like slender necks bent in deference. The fruits and the earth connect. The stems keep bending. They push down until the capsule is buried in the soft moss. The plant, *Spigelia genuflexa*, has planted its own seeds.

A handyman and plant collector named José Carlos Mendes Santos, known better as Louro, discovered the plant as a new species in 2009 while squatting behind a bush for a "common human activity." He was near the aforementioned off-grid house of Alex Popovkin, an amateur botanist for whom Louro often worked. The pair found a few more individuals nearby and watched them move through their life cycle for two seasons. After researchers in the United States confirmed it was a new species, the pair copublished their discovery in a peer-reviewed journal. The plants, they wrote, would emerge in the same spot each March, exactly where their parents had planted them. Whereas birds shelter their young in nests and small mammals dig burrows, *Spigelia genuflexa* plants their babies in a bed of moss, the safest, best place to ride out a months-long dry season.

Botanists have long known that plant parents will go to great lengths to give their young a better start in life. In this case, the Brazilian plant better ensures the success of its child by deciding exactly where it should germinate. And in a harsh and variable landscape, the best spot is the one already proven to be fertile; the parent has already managed to grow there. Even the botanists most squeamish about using other colloquial terms for plant behavior (including the word "behavior" itself) would call what the Brazilian plant does a demonstration of "maternal care." I find this a funny bit of poetic license. Unless you live on a small island or inside a ginkgo grove, the majority of plants you encounter are bisexual, meaning they have male and female parts, and are capable of producing the plant equivalent of both eggs and sperm. Indeed, the Brazilian plant is capable of "selfing," meaning that, like many plants, it will sometimes combine its own pollen and ovules to produce offspring. "Parental care" might be a more accurate moniker, unless you are willing to take a more nuanced view of the fluidity of plant sexuality; indeed, when the plant is dealing with its own fertilized ovule, one could say it is busying itself with the maternal part of its life phase. I like to imagine bisexual plants as something like the androgynous beings from Ursula K. Le Guin's novel *The Left Hand of Darkness*, who are capable of both siring and gestating children at turns, being a mother in some instances and a father in others. They quite pity the human visitor who can only do one.

Maternal care in plants—to use the prevailing term—is widespread, though the way that spigelia bends down to plant its own fruit is quite rare, though it is shared by the common peanut. There are many other ways that plants care for their young. Whereas small mammals huddle over their young and lizard and snake parents bask in the sun and then drape themselves over their eggs, transferring their body heat, plants also carefully adjust the temperature of their developing embryos. The narrowleaf plantain, an extremely common edible weed in parks and lawns and cracks in the sidewalk, grows its seeds on a tall exposed spike. When air temperatures are high, the plantain lightens the color of the spike, and it darkens it when the air is cool, to reflect or absorb the sun's rays as needed to keep the developing seeds at an ideal temperature. Many plants alter the thickness of the fruit wall and of the seed's protective coat—both of which are actually maternal tissues—to adjust the timing of the seedling's emergence. If a parent plant finds itself in a drier environment, it may make seeds with a larger surface area, so more water can pass through the seed's porous surface, keeping the embryo within well hydrated. In the high alpine ridges of Colorado, some plants are known to deposit their seeds directly at the base of their own stem, much like the Brazilian moss-dwelling flower. That way, the baby plant can start its life in the shade of its parent, in an otherwise exposed and sun-bleached landscape where a young plant is liable to dry out into a botanical bonito flake within days. As the parent plant dies, the moisture in its decaying body goes to nourish its child too.

But there's another way that plant parents set their children up for success. They pass on the wisdom of their experiences. New research is resurrecting an old idea; that the environment in which a plant lives is inextricable from the plant itself. The environment informs what kind of plant the offspring will become; the kind that survives and thrives in a challenging scenario or doesn't. It changes their body plan, possibly directs their development. And those changes can be passed down to their offspring, whose body will develop differently from the start, and be better able to handle the harsh conditions its parents experienced.

In other words, parent plants can pass on skills for surviving in a tough world. In some cases this involves whole new body parts and coats of armor. For example, if yellow monkey flowers are exposed to predators, they will produce babies with a quiver of defensive spikes on their leaves. Wild radishes that have lived through a scourge of destructive caterpillars will make baby radishes with extra-bristly leaves too, plus they'll be preloaded with defensive chemicals to better ward off threats. If these plant-children end up facing the same challenges their parents did, they'll be much better prepared to handle them.

These changes can be dramatic, the kind of thing that science would assume is the purview of hardwired genetics, which are themselves the result of evolution. But it all happens far too fast for that. No plant evolves in a single generation. Genes, it seems, don't tell the whole story. They may not even tell half of it.

In a previous chapter, we learned about plant memories, the way plants can recall their own past experiences to make informed choices and change their trajectories. But what about generational memory, the kind you inherit? Now that researchers are starting to look for them, these transgenerational effects are threatening to change the entire field of plant genetics, or "evo-devo," the study of evolutionary development. In response, "eco-devo," or ecological development, has sprung up as a new discipline to study the massive influence of the environment. Genes are the current stand-in for the code to life. Genes are of course import-ant to many things in a plant's life. But increasingly it seems like they're less akin to a code that the organism reads out than to a flexible reper-toire, a choose-your-adventure novel with a multiplicity of endings, each influenced by a million subtle changes in the storyline.

If genes don't tell the whole story of what a plant will become, a new theory of life needs to fill in the gap. Plants have a wide range of flexibility to transform into what their surroundings demand of them. Every aspect of the environment a plant experiences—and the environment its parents experienced—may play a bigger role in shap-ing plants' being than anyone thought. Or put another way, plants are shaping their own futures. They're adapting their bodies to better suit their changing environment. The environment is working on them, and they are working on themselves in response, sculpting themselves into new kinds of plants.* According to Sonia Sultan, a plant evolution-

---

* This is seen in animals too: when brown frog tadpoles develop in the presence of

ary ecologist at Wesleyan University in Connecticut, this means plants have agency. And by passing adaptations to their babies, plants are directing the trajectory of their species. They may be more in charge of themselves than anyone thought.

When I find Sultan in her office at Wesleyan in midsummer, we get to talking about her childhood. Sultan was raised in Massachusetts by two New Yorkers. Her father was an English professor, and her mother is a psychologist. Her first word was "flower," or so her parents tell her. It was the first thing she thought really needed a name. She spent her early childhood in the university greenhouse, where she wandered the rows of plants and played in its climate-controlled corners. She developed her affinity for plant companionship there. She came to think of plants as quietly capable, doing what must be done and doing it well, without fuss, all from the confines of their little pots. They gave her a sense that everything was being well handled. Nothing about that has changed for her. "I like being around them," she says. "They have this serene competence."

Sultan's mother, the psychologist, didn't understand why she chose to study plants and not people. "She thought that was like a personal diss," Sultan says. Yet in her approach to studying plants, Sultan may well be changing how science thinks of the trajectory of all life, people included.* Her conclusions might apply just as much to a hyacinth as to homo sapiens. Our environment, she keeps writing, in a career's worth of scientific papers, is inextricable from who we become, and who our children become. And that fact might just prove that plants—and every

---

a predatory salamander, they develop a bulging body that is too large for the salamander to eat. The red knot, a shorebird, can develop a larger pectoral flight muscle within a few days of the presence of a predator, allowing it to more quickly escape. These examples were gleaned from Sonia Sultan's 2013 lecture, "Nature AND Nurture: An Interactive View of Genes and Environment," at the Institute for Advanced Study in Berlin, https://vimeo.com/67641223.

* Plants are just more useful for this sort of research because they can more easily be cloned.

other living thing—have agency over their own development. They take into account their conditions, and then shape their own structure and function accordingly. At a deep, biological level, of course. No one is suggesting that plants are willing themselves a new set of leaf spikes.

The position made Sultan wary of being misunderstood. At first, she didn't want to talk to me. She'd rather not be included in this book at all if I was going to lump her in with the "plant intelligence" people. Journalism on these sorts of plant science ideas hadn't so far had a great track record of capturing nuance, or thinking beyond human tropes, and for her, it was a matter of academic life or death. She hadn't bothered trying to get an NSF grant in twenty years, like many of the plant researchers I'd talked to so far, but she still had peer reviewers at journals to contend with. The conflict in her field, she said, was "heating up." She's felt a bit under siege. I told her I was interested in nuance; I understood that she wasn't saying plants had brains, or that they had the capacity to think like us. Agency here means something else, something more basic to all life. That, to me, felt just as enchanting.

Sultan has short dark hair and calcite-blue eyes. She pauses when speaking to select the best word. She is quietly serious but also slyly funny; she tells me she considers the main problem with humanity to be that we are more akin to chimpanzees than to bonobos.* Her office door in the biology building is covered in cheeky lab jokes and ephemera, some clearly concocted by her students. There's a picture of a bereft-looking seedling with a speech bubble that says "slipping into darkness"—much of her greenhouse work is about how plants grown in shade go on to produce offspring already equipped to grow better in shade; thousands of plants have been subjected to dim conditions on her watch. There's a printed excerpt from the novel *Good Omens*, by Neil Gaiman and Terry Pratchett, about a man who decides he ought to verbally berate his houseplants into growing better. It is labeled "New Sultan Greenhouse Protocol." I notice a ceramic mug in her office made

---

* Bonobos have matriarchal societies.

to mimic the Hellenistic takeout coffee cups one gets from diners and carts in New York City, the kind that say "It's our pleasure to serve you." Sultan says that people tell her she seems like she's from New York. Sometimes you can't help but pick up the markers of your parents' home environment, I think.

In high school Sultan took a forestry course that made her aware of plants as individual species, each with their own names and quirks. She finished high school early to take an internship at Arnold Arboretum, at Harvard University. She found she liked being around plant people; they were the rare kind who knew all about the comfort of plant competence that she'd absorbed so young. In her undergrad years at Princeton University, she studied history and philosophy of science, which cemented in her the awareness that science is not objective, and scientific paradigms come and go, each with their own blind spots and biases. "Science is not the objective accumulation of facts," she says. "The ways of thinking that scientists use are things that scientists invent."

When an old paradigm falls away, in preference of a new one, everyone acts as though they'd known the new one was the truth all along. These changes have major ripple effects in virtually every discipline, Sultan tells me, almost as soon as I've arrived. Copernicus's discovery that Earth and the planets rotate around the sun served as the inspiration for William Harvey's discovery of the circulatory system, after all: "He saw the heart as the sun in the center of the body." What would have happened without that? We inherit knowledge from previous generations. A science built on a faulty premise might lead to a number of incorrect assumptions; scientific discovery builds on itself. If there is a hairline crack in the foundation, the fissure will spread through everything built on top. The structure will not hold.

Enter the genetics revolution. Sultan sees it as both a foundation and the ominous crack within it. It's not that the discovery of genome sequencing was somehow incorrect—it was a remarkable leap for science, and brought many incredible discoveries that expanded human knowledge of how life works. But she's more worried about the vise grip it now has on science's approach to questions that remain unanswered, of which there are so many. And on practically all the scientific funding.

The observation that a plant will develop very differently in one environment versus another is unavoidable to anyone who studies plants. "It tormented the scientists of the twentieth century," she says. If they paid too much attention to it, it would surely ruin the results of uncountable experiments. Any variation was instead considered a quirk of a specific individual, an outlier in the data. There were a lot of outliers. But the idea that plants might be governed by more than just their genes would puncture the sheen of accomplishment forged by the genetic discoveries of midcentury. Scientists had found the basis for life. In the all-or-nothing thinking to which Western science tends to be devoted, the new genetic paradigm was absorbed wholly, and left no room for this kind of ambiguity. So it was mostly ignored.

Genes were the puzzle pieces that made every living thing, and if we could find what each piece did, we'd know everything about organisms. They would become fully predictable. This belief extended far beyond plants, of course. Human genetics took on a godlike omnipotence. The gene for intelligence, the gene for homosexuality, genes for diseases and psychological conditions, they were all waiting to be found. For example, the race to find the gene for schizophrenia absorbed millions of dollars and the whole of many professional careers in the early decades of genomics. It seemed to be heritable, but not always, and it didn't work the way traditional Mendelian genetics suggested it should. They never did find the schizophrenia gene, but the quest to find it trundles on to this day.*

"The DNA sequence ... spells out the exact instructions required to create a particular organism with its own unique traits," reads a para-

---

* Of course I mean "suite of genetic markers," not just one gene. And researchers have found genetic markers that seem to increase a person's risk of developing schizophrenia, but no genetic answer has yet been the key to explaining who gets it. For more on the mystery of schizophrenia, see the masterful book *Hidden Valley Road*, by Robert Kolker (New York: Doubleday, 2020).

graph on a U. S. government website about the human genome. To Sultan, that sums up the problem. Genes are not exact instructions. They're more like stage cues at an improv show. A lot of other things could happen along the way.

That's the irony: Mendelian genetics were never universal. Really, they're a "boutique subset" of how genes get combined and transmitted through generations. "The big A and the little a, the gene for tall and the gene for short. That's really not how the vast majority of genes work," Sultan once explained. In fact, genetic inheritance seems to explain only about 36 percent of the heritability of a person's height, one of the physical traits that appears to be the most reliably related to the physical traits of your parents. Scientists call this perplexing phenomenon "missing heritability." No one yet knows what fills the gap. "When will people like me stop teaching Mendelism as our model for genetics?" Sultan asked.

It all reminds me a bit of Descartes, and his view that animals were machines that could be disassembled and reassembled if we just knew all the parts. The idea of genes suggests little parts in a machine too; proteins and receptors that coded for specific outcomes. It's a mechanistic view of life.

This was the scientific environment that Sultan came up in. It was exciting; new discoveries appeared within reach inside the genome for anyone who could pose it a good question. Genetics were the long-lost key to life, and now anyone graduating in her profession was expected to start using it on as many locks as possible. When Sultan arrived in graduate school at Harvard in the 1980s as a brand-new population biologist, she started trying to show that plants in sunny environments had sunny-plant genes, and shade plants had shade genes. In other words, they were genetically predisposed to be in those places. Everything she'd learned had prepared her to view the world that way. But every morning on her walk to the bio labs, she looked at the plants in their manicured plots around campus. They seemed to contradict her lab work. In the real world, the same species of plant looked very different if it was growing in a sunny patch or a shady patch, or in a crack in the sidewalk versus an unobstructed plot of soil. The shape and size of their leaves, their stoutness or tallness, their overall look, were thought to be genetically determined. But how could this variation in the same species be part of a plant's genetic code? Genes couldn't control where a seed landed, and evolution didn't happen that fast. Something about the environment itself seemed to be changing the very shape of these plants.

That made her wonder, if that's true, then development might be much more complicated—and more interesting—than we think. What she had been taught were settled principles were instead open questions. For the last thirty-five years, that's what she's studied. In one experiment, she found that a plant's body size can double or triple if grown in low light—more surface area to catch falling photons. Plants grown in an excess of water, meanwhile, will change their bodies to avoid drowning, producing unique hairlike roots at the very surface of the soil, to access oxygen even when the soil is waterlogged. I thought about goldfish, the kind children get at fairs, which can completely remodel their gills to increase respiratory surface area after a few days in low-oxygen water. Bigger gills mean a better chance of catching some oxygen. Sultan had found the goldfish of the plant world.

But when you deprive a plant of water, it will grow less tissue in total. This one is fairly logical: if a person doesn't have enough to eat, they will also grow less body mass. If the deprivation is bad enough, their growth will be stunted. But water-deprived plants, hoping to improve their lot, will use what little body mass they can muster to maximize the surface area of their roots. They put more of their limited tissue into their underground parts, growing them extra long to search for water, but extra thin to cover as much ground as possible on their limited diet.

Roots, we've already learned, are foraging organs. In some of Sultan's early experiments, she moved water to different spots in soil and watched her plants' roots follow it around like dogs on a scent trail. But her more recent work looks at whether drought-stressed plants will have different sorts of kids than plants that grew up with all the water they needed. She found that when a plant is grown in dry soil, and then reproduces, its babies quickly develop a body that's expertly suited to drought when they find themselves in dry soil too. They don't hesitate; this second generation morphs into deep-rooted, long-rooted seedlings. So does the third.

In an empty classroom next to her office, Sultan drew a graph for me on the board. It's from a study from 2000 on smoking, genes, broccoli, and lung cancer. People with the "lung cancer gene" are much more likely to develop lung cancer, especially if they smoke. Due to genetic mutations, they lack a certain enzyme that most people have, which normally works to clear cancer-causing chemicals from hazards like tobacco smoke out of the lungs. So that was one trend line, at a steep angle upward; the more these "lung cancer gene" people smoked, the more likely they were to get lung cancer. But then she drew a second line and labeled it "broccoli." People with the lung cancer gene who ate a significant amount of cruciferous vegetables, including broccoli, appeared to have a lower risk of lung cancer proportional to how much broccoli they ate. At high enough intake, broccoli nearly erased the impact of the gene mutation. (In a different study, it even appeared to help people without the genetic anomaly clear carcinogens from smoking.) This is likely because cruciferous vegetables like broccoli produce compounds that, when broken down in our bodies, turn into the enzymes that detoxify cancer-causing chemicals. In other words, these vegetables make the thing that does what the "lung cancer gene" people lacked the genetic ability to do. Here, genes weren't everything. The environment a person experienced—in this case, what they ate—had a place in the story too. Possibly even a bigger place than genes.

"And you wonder why every medical researcher isn't going after this," Sultan says, rather than the endless hunt for genetic causes. Some are, of course. Just not as many as she thinks should.

Next to the broccoli graph on the board, she drew a graph of leaf size in plants, showing that the less sun a plant got, the bigger its leaves would get, straining to catch more light. Same thing, she said. The environment dramatically changes a creature, be it a human or a plant. "Biology is biology."

This means everything a plant experiences changes its outcome. No environment is neutral. Even the supposedly "standard" form for any given plant is likely the impact of its environment. This of course confounds a lot of lab work.* Sultan says she loves the realization that reliably flickers across the face of her students at some point in their course of study with her. *Wait*, they say. *This means there is no control environment.*†

The environment appears to flow through organisms, altering them at the deepest level. In a 2015 book, Sultan wrote that this makes it hard to even consider the two as fully separate entities: "On closer examination, however, the environment extends into the organism and the organism into its environment, in ways that obscure the boundary between them." The influence flows in both directions, she wrote—the organism shapes its environment while its environment shapes it. The figurative membrane that separates organisms from the rest of the world doesn't just leak; it lets the rain all the way in.

Consider the emerald green sea slug. When I first read about it, I couldn't stop talking about it with anyone who asked what was going on with me lately. The green sea slug, this whimsical thing that seemed to defy all boundaries between plant and animal, was going on with me. It was all I could think about.

The slug, which lives in watery places all along the Atlantic coast

---

* When biologists started studying development, it was obvious to them that studying organisms in the context of their environment was the only way to study their development. It was only in the mid-twentieth century that scientists began removing organisms from their natural environment to study them in the artificial, "neutral" context of the lab.

† With that in mind, Sultan tries to make her greenhouse as close to the outdoors as possible. She only does experiments in the summer, with real summer sun, and uses a soil mix with a more "naturalistic" texture, "because I want to see roots grow the way they would." The greenhouse is like an intermediate zone between a completely artificial lab and the real world. "I'm probably the only person who has a greenhouse where we use clay pots," she says. "Not out of a romantic reason, but because the clay breathes. So it's more naturalistic than plastic."

of the United States, spends its early life a brownish color with a few red dots. It has one goal in those early moments: to locate the hair-like strands of the green algae *Vaucheria litorea*. When it finds them, it punctures the alga's wall and begins to slurp out its cells as though through a straw, leaving the clear tube of empty algae behind. The algal cells are bright green, owing to the chlorophyll-filled chloroplasts inside them, which are responsible for photosynthesis. Under a microscope the whole exchange looks uncannily like the slug is drinking bubble tea, one bright green boba entering its mouth at a time. The slug digests the cells but keeps the chloroplasts within them intact, spreading them out through its branched gut. Now the slug itself has turned from brown to a brilliant green. After a few algal bubble teas, the slug never needs to eat again. It begins to photosynthesize. It gets all the energy it needs from the sun, having somehow also acquired the genetic ability to run the chloroplasts, eating light, exactly like a plant. How this is possible is still unknown. Remarkably, the now-emerald-green slug is shaped exactly like a leaf, all but for its snail-like head. Its body is flat and broad and heart-shaped, and pointed at its tail end like a leaf tip. A web of leaflike veins branch across its surface. The slug orients its body in the same way a leaf does, angling its flat surface to maximize the sunlight that falls upon it.

The green sea slug blurs the boundary between animal and plant. But it also offers a striking example of how easily the boundary between organism and environment can be traversed. The slug's essence is acquired through an interaction with its environment. The algae is part of the slug's environment, and the slug is quite literally transformed by taking it in. There couldn't be a leakier creature. It's an extreme example, of course, but a version of it happens to us, and our bodies, all the time. We are constantly taking our environment in, and it is constantly transforming us. We wouldn't be us without it.

We are drawn closer to plants in this way. Our outcomes are massively entwined with our environments. What we eat, the air we breathe, and the various substances we are exposed to all have the power to change the direction of our lives and bodies. We think of our own development as being scripted by the genes we inherit. But our in-born selves also include this environmental message from our parents' environment, and in many cases even more distant environments in our family line, Sultan says.

All biology, I began to understand, is in fact ecology. Ecosystem dynamics that ecologists study apply just as easily to single plants themselves. Resources like food and water fluctuate in an ecosystem, which causes different individuals to take up residence in different groupings at different times. The traits of the community change, based on the changing environment. But if you zoom in on a single plant, the same is true. The environment influences the individual, too. The various traits an individual could have are in constant flux in response to how the environment changes, much like how the various actors in an ecosystem fluctuate whenever anything changes about the environment they lived in.

The Italian philosopher Emanuele Coccia wrote that plants exist in a state of total "immersion." Immersion is an action of "compenetration," he wrote, a word that means pervasive, mutual interfusion. It seemed like the perfect word for everything I'd learned so far, and I thought of it often as I learned more and more about the world of plant development. In plants, "*to act* and *to be acted upon* are formally indistinguishable," Coccia wrote. "If the environment does not begin beyond the skin of the living being, this is because the world is already inside it." To exist, for plants, means to reciprocally construct the world. The world is inside them. There is no other way.

Plants are particular exemplars of this sort of immersion, in part because they cannot get up and move to adapt to changes in their environment. They can't run away. So their porousness is extreme, exaggerated. We can run from threats. We can physically move our bodies to more suitable conditions. But our own immersion in the environment is total also, in subtler ways. What plants illustrate best is that none of us can actually run from the impact of our environment, and the environment of our parents. It's already inside us. Suddenly I was seeing a world of fluctuating, interpenetrating elements, our bodies open to them all. Just living in the world exposes us to so much; all of it, cumulatively, makes us us. The idea that everything is interconnected came through like a powerful chime. Everything is quite literally interconnected. The evidence is us.

I found clear examples of this in my life as an environment reporter. A few years prior, I'd gone to Detroit to interview people living in a neighborhood surrounded by oil refineries and coal-fired power plants and garbage incineration facilities. The rates of asthma and other respiratory illnesses were staggering there. This made sense; the air was clearly unsafe to breathe. I was soon told that babies in the area were regularly born with asthma, and doctors sometimes gave parents of newborns nebulizers to take home with them. I learned how the particles of air pollution a pregnant person breathes can make their way into the bloodstream, slipping into blood cells that flow to the developing fetus, delaying and damaging its lung development. Babies are born prepolluted. But then I learned something more startling: I was on the phone with Kari Nadeau, a physician and researcher at Stanford University who studied the way exposure to air pollution ripples across generations. She told me those same pollution molecules that cross a placenta can also change people who aren't pregnant at all. They can slip into the blood that feeds ovaries and testicles, altering genetic expression. If those are altered, so are any offspring created by the eggs and sperm those organs might go on to produce. In fact, Nadeau was able to infer that the genes of her patients—who lived in Fresno, in California's Central Valley, the most polluted city in the state, due to a deadly combination of diesel exhaust and agricultural pesticides—were fundamentally altered so that they would be more likely to develop asthma and allergies. And those genetic changes could be passed down to their children, and their children's children, even if those later generations had moved away and were no longer exposed to the pollution.

This is a very bleak example of human epigenetics—the ways our specific environment can modify the way our genes work, and how those changes can be passed to our offspring, their offspring, and so on. But there are likely to be scores of other examples that help fill in the gaps in our understanding of the fundamentals of our lives. Perhaps one day we will close the "inheritance gap" on why kids tend to grow to a similar height as their parents, for example. Meanwhile, dozens of diseases that seem to strongly travel in families fall under the missing heritability problem too, like Type 2 diabetes (only 6 percent seems explained by inherited genes), early-onset heart attacks (less than 3 percent), lupus (15 percent), and Crohn's disease (20 percent). Perhaps it will have everything to do with their environment, and their parents' environment, and so on.

I thought, too, about Ernesto Gianoli's theory of microbial infection to explain how the boquila vine in Chile managed to mimic all those species. A change in their microbial worlds, he thinks, caused a change in their very shape, a thing we think of as fundamental to a species, its very essence. Perhaps our "essence" was more flexible than we thought. Perhaps it was contiguous with, and not separate from, our environment. Gianoli's theory, whether or not it turns out to be true, only piggybacked on a far more established yet still recent revelation: each living being, be it a plant or a fish or a human, is fully infiltrated by legions of microbes. These microbes often have yet-smaller microbes living inside them. Each of them is subject to environmental changes, too. Were they not a sort of community in themselves, and the body they lived in an ecosystem? It makes sense then, when we scale up to imagine a whole individual plant, or person, to not lose sight of its fundamental architecture, which is truly that of a community of creatures responding to changes in their world. Everything, at every level of life from a microbe to a rain forest, then, is an ecosystem. We are more like a system than a single unit. All biology is ecology.

Plants remind us that we are contiguous with our environment, impacted by its every fluctuation, impacts that reverberate through our lineage. Our environment shapes our lives and the lives of our descendants. We inherit their environment in bodily form. One could say we inherit the earth.

OF COURSE, THERE are limits to how plastic a living thing can be. Not all changes are surmountable. Take a wildfire, for example. No plant that is not already evolutionarily adapted to fire will spring into action and become suddenly fire-resistant. And plasticity likely varies dramatically between species; sometimes it will depend on where they evolved. Some, like the many "naive" native plants of Hawai'i, have very little capacity to adapt to change, having evolved in a place without natural predators, and are handily outwitted by invasives. They just aren't that plastic.

But other plants are fantastically plastic; their plasticity seems to know no bounds. New environments embolden them to new shapes. Enter the invasive species, the undefeated stars of plant agency.

They're super plastic, and extremely adept at passing that plasticity on to their children. "I admire their capacity, biologically," Sultan says. There's an assumption in biology that says that natural selection, in general, creates species that are highly specialized. They're very good at what they do—say, growing in one specific habitat—and they're pretty bad at doing anything else. Other species might be generalists, capable of surviving in more places, but they're not particularly good at anything—they survive, but don't exactly thrive. "The jack of all trades has to be a master of none," Sultan says. In life there are trade-offs, or so says the conventional wisdom. But some invasive species scoff at the concept. "They're good at everything," she said. Jacks of all trades, and masters of all. "That's not supposed to be possible."

Sultan studies *Polygonum cespitosum*, whose common name is smartweed, if you can believe it (I could not). Sultan tells me the origin of the name is actually the fact that it produces an acid that will smart if it gets in your eyes. Still a fantastic name, I think.

Smartweed was first imported from Asia, and has become terrifically invasive in northeastern North America. It appears to be the prototypical weed. Not showy, nothing particularly striking about them. She likes that they are an utterly ordinary weed. Plus, they're easy to clone. It helps to have lots of genetically identical plants when you want to see what happens when everything other than genetics changes.

On Sultan's lab website, it says the team studies plant "monsters." These are monsters, I come to understand, of their own making. "It's a term of admiration," Sultan says. *Polygonum cespitosum* has managed to evolve very rapidly to suit their new environment. They've become capable of colonizing all sorts of habitats in their new North American home. They have developed rapid lifecycles, during which they are wildly successful at reproducing. The ones who are best at this will no doubt become the future of their species, fueling generations that are themselves faster and more successful at reproducing. The result? Smartweed is rapidly evolving toward greater and greater invasiveness.

Only about one out of a hundred plants that get introduced to a new place become invasive, defined most often as a non-native plant that spreads quickly and has the potential to cause harm, either ecological or economic. Plants and animals are being introduced into new places constantly. Most of them just blink out. They don't have something in this new place that they really need, like a specific pollinator or a certain temperature range. They just don't take. But a subset of these species will persist. Perhaps the temperature range in this new place is similar to the one they evolved to live in at home. Maybe they are not too particular about what type of creature pollinates them. They hang on.

Of those species that hang on in their new home, a very small portion of those actually fare better even than the native species. They crowd out the previous inhabitants and expand their range. There is generally a very long lag time, about fifty or a hundred years, between the time the plant arrives and when it suddenly goes gangbusters. All of a sudden people start seeing it everywhere. *Polygonum cespitosum*, for example, was declared an invasive in the early 2000s. It was likely introduced in the early part of the 1900s. Why this lag? It indicates, to Sultan, that the plant didn't necessarily arrive with the skills it needed. It didn't show up and immediately take over. What was happening during those decades? It may have been rapidly evolving—thanks to certain extraordinarily plastic individuals who were particularly good at passing down crucial environmental skills that they learned on the fly, changing their bodies accordingly. It's a different way to look at invasion biology. It's not that the species as a whole was necessarily so invasive. But certain individuals were just so flexible that they could fine-tune their bodies to suit their new homes, and were also so good at passing that flexibility to their offspring, that the species as a whole transformed into the perfect plant to exploit their new conditions. It takes time to learn a new place, biologically speaking.

Of course, not all individual smartweeds are the stars of this show. Through careful greenhouse studies, Sultan has found that, as is to be expected, each individual smartweed reacts a bit differently to changes in its environment. But some are prodigies of adaptation, elite athletes of plasticity. They may prefer wet environments, but will do just fine in dry, adeptly switching gears away from growing leaves to growing longer, thinner roots to seek out any scrap of moisture. They may like full sun, but will make do in the shade, growing larger leaves. Most remarkably of all, they pass these adaptations on to their children. The offspring of one of these adaptable plants grown in drought-dry soil will adapt even faster than its parent when faced with dry soils themselves, quickly springing into action to grow long thin roots. It already knows what to do. Similarly, she found that smartweed plants whose parents had to compete against neighbors for light will grow bigger leaves, ready to outcompete their neighbors for space in the sun. Plus, if a smartweed plant is grown in the shade, and reproduces, its babies quickly develop a body that's expertly suited to do well in shade, growing taller than their parents to reach more light, and larger leaves to catch more of it. They also flower sooner, suggesting better reproductive fitness. What's more, when the offspring of shade-grown parents are pitted against neighbors whose parents were grown in the sun, and everyone is made to grow in shade, the offspring of shade-grown parents handily outcompete them for space. These seedlings have a type of generational wealth: they've inherited a useful skill. They will have a leg up on their peers when the same hardship their parents encountered befalls them.

These individuals are the future of their species. They will survive better, and reproduce more. Their offspring will survive better too, likely even better than they do, equipped with this inherited plasticity, like a superpower that lets them flourish even in tough times. Invasive species have often been maligned as more aggressive, ruthlessly competitive. These are strangely moralizing concepts to put on a plant, when you think about it. The words we use for invasive species are very often unsubtle in their xenophobia, matching nativist language. We call them "aliens" and drape tropes on them about being unnatural in their abilities, aggressive by nature, like diseases on the land. But what if they're just more resourceful, more plastic, better at handling change and passing on wisdom to their young? They certainly disturb our landscapes, and supplant species that we have grown to love, in our short evolutionary stint on the planet so far. But we are the youngest brothers of this family of ours. We haven't been here that long. Change happens, plant communities fluctuate. Of course, there's a twist. This era is unique in that *we* are the ones moving plants around the world. We've caused—and are still causing!—most of these invasive species to show up in new places, to adapt to new scenarios and locations. We literally bring them there. It's even stranger to fault a plant for its successes with this fact in mind.

Take Japanese knotweed, for example. There may be no plant more successful on the planet, or more hated. It is a close relative of the smartweed Sultan studies. Japanese knotweed was first introduced to North America in the 1860s by plant collectors seeking to add an attractive exotic species to their private nurseries. It had already been imported to Europe several years before. The plant became popular for its white flowers and dense coverage; it grew preposterously fast and thick as a hedge, perfect for privacy along roadsides. Apparently some people in the United States still intentionally plant it in their gardens. They clearly have no idea what they're about to face. Once you meet a knotweed, the fact that plants can morph their bodies at will becomes searingly obvious. Their agency is almost palpable. Any illusion of human control falls away. I met it for the first time one day in late April.

I came out onto my porch after a drenching rain to find new spikes of fleshy reddish-green plants poking up from the sides of the yard, right up against the wooden fence. We'd just begun subletting a friend's apartment, and I was giddy with the sense of potential of having a backyard for a growing season, an unfathomable luxury in New York. Inside, on the windowsill, we sprouted seedlings of hot peppers and radish and mustard greens in shallow cardboard boxes, waiting for the still-icy weather to warm enough to plant them in a raised bed in the yard. We watered them a trickle at a time, squeezed from a plastic water bottle with a perforated cap. The vegetable seedlings exuded fragility. Meanwhile, outside, the reddish shoots exuded the opposite. They were robust, authoritative, clearly undaunted by the chill or the driving rain or especially not the heavy black tarp that underlaid the wood chips in the yard, placed there specifically to repress them. No bother, I imagined the knotweed saying to the tarp.

I started pulling it out. The hollow stems broke off at their base with a juicy pop. I'd read that their shoots were edible and nutritious at this time of year, tasting something like a cross between rhubarb and sorrel, but I glanced at the pile of construction debris in the adjoining yard—itself studded with knotweed shoots even taller than mine—and decided there were too many chemical unknowns around here to risk eating it.

Two days later I came back again to find new shoots, already several inches tall, just next to the places I'd pulled their brethren out forty-eight hours prior. They hadn't been there at all on my first foray. While my vegetables limped along on their windowsill, growing slower than seemed reasonable for all the pampering I gave them, the knotweed was raising itself at the speed of an inflatable pool.

I read that knotweed grows rhizomatically, with continuously growing underground stems called rhizomes. Once established, it spreads a complex network of rhizomes beneath the soil, a subway system of runners and shoots, a continuous network with no clear center. It's almost impossible to dig out and remove these vast rhizome networks, and without doing that, the plant can never be extricated. A fingernail-sized root fragment left behind can regenerate the entire plant. And knotweed-blanketed areas the size of half a football field have been found to be a single individual, one gigantic rhizome monster. My plucking was futile. I looked over the fence. Our shoots were likely the sentinels of the plants back there, threading their runners beneath the crumbling pavers on that side and the black tarp on this one, popping out at any small crack or seam or tear, or perhaps making their own, I didn't know.

By May the knotweed had infiltrated my planter box and grown halfway up the five-foot fence. Just over the barrier, in the abandoned yard of the building under construction, the knotweed had completely taken over. By June the neighboring yard was a knotweed forest, the bushes dense and tall as me, waving in the breeze over the edge of our shared fence. Sitting in the yard I had the impression of sitting in a clearing in a jungle. I had to admit that the knotweed was beautiful. Its bright green leaves were round and broad as my palm. Their thick, juicy stalks, speckled with red, gave the impression of total good health. As I walked around my neighborhood all that month, I saw whole empty lots between buildings transformed into knotweed utopias from chain-link fence to chain-link fence. I considered, with a thin sense of fear, that I was only seeing half the action. What invisible vegetal architecture was spooling out belowground? How soon until those rhizomes ran out of room in their cozy lot and began poking into the foundations of the row houses on either side?

There's no doubt that knotweed can burst through pavement by seeking out and exploiting cracks in foundations, infiltrating the fissure and widening it to suit its needs. If a niche is unavailable, it quite literally creates one for itself. That niche just happens to be the one we hoped would be impenetrable: the niche we built for ourselves.

Even just a single tendril of green flesh rupturing concrete begins to split open our understanding of plants as sessile, squishable, inert. A soft thing without eyes or a mouth that applies sustained pressure to our hard boundaries, the only thing between us and the dirt, and it wins the fight? That's unsettling to a sense of order bent on the supremacy of human ingenuity. The possibility that we are not in charge flicks through the mind. Power is a matter of perspective.

It's under some dispute just how much damage knotweed can do to a house, though its tendrils have been found making their way through walls, emerging indoors. Already in the UK, the mere presence of Japanese knotweed on a property makes it unmortgageable. The country made knotweed disclosure mandatory on all deeds of sale, and banks won't issue a mortgage to a property with knotweed on its grounds or growing within three meters of it unless the owners have a management plan, which, considering how much earth must be moved to extirpate a knotweed rhizome network, is financially impossible for most.

In the United States, land managers at the National Park Service anxiously watch knotweed grow ten feet in a single season, springing up so quickly that herbivores like deer stand no chance of meaningfully nibbling it to death or trampling it before it matures, as they can do with a great number of native species. At Acadia National Park in Maine, troops of volunteers fan out into the forest, cutting the stalks and then applying herbicide to the stumps, in the hopes that the plant sucks it into its rhizome, killing itself systemically.

Japanese knotweed is now one of the most invasive plants in the world—or one of the most successful, from the plant's perspective. It is thought to flourish on every continent besides Antarctica. In New York City, the *New York Times* reported that there were currently miles upon unbroken miles of the plant growing along the Bronx River, along the Hudson River, and in all five of the city's boroughs. It is, in no uncertain terms, here to stay. Its future is secure, thanks to its incredible plasticity, the agency it exudes through every new growing tip. We brought it here. It's just doing a great job at being a plant where we put it.

There are several ways Sultan's research could prove to be useful in the world. If we understand how species become invasive, we might be better able to predict which ones will take off by analyzing their potential plasticity. Plus, we know we're changing the planet faster than many plants can evolve to keep up. But if we can better understand what makes certain plants more plastic than others, we could potentially help organisms exist through climate change, "and all the other shit we're throwing at them," Sultan says. We already know how many of these changes will take place: where it will get hotter and drier, or hotter and wetter. One could see a future where we can identify the most plastic genotype of any given species for any given outcome, for example, and plant those out, giving a vulnerable population a boost.

The idea that plants have agency is bubbling up through the literature now, with Sultan at the helm (Simon Gilroy and Tony Trewavas are up there too). *Agency* is an emotionally loaded word. Sultan is taking a gamble by using it. It immediately calls to mind the existence of a mind, of intention and desire. But she says we need to get past that. "It's not intentional, and it's not intelligent in the way most people use that word. But it is agency," she says, clearly trying to distance herself as much as possible from anyone trying to portray plants as little humans. Agency is an organism's capacity to assess the conditions it finds itself in, and change itself to suit them. Yes, we do this all the time. So do plants.

Mechanistic visions of being controlled only by our preprogrammed genes do not satisfy our innate understanding of ourselves as brimming with irreducible complexity and subtlety. We too are like plants, taking in information from outside our skin, the membrane separating us only barely from the world we live in. Beneath the surface of every organism there is a vibrancy we do not know, yet, how to completely dissect or replicate. Complexity is mounting instead of receding. And that's okay. It's probably the truer direction. It just might bring us new revelations about all life.

The conversation turns to Sultan's own garden, at her home nearby. She grows garlic and herbs and a few vegetables, but those are rather unsuccessful, because she can't bring herself to use herbicides or to pull up many weeds. "My backyard is full of weeds, in the sense that whatever thrives there I feel has a right to be there," she says. "They want to grow. This is what they do. Who am I to be constantly stomping on them, whacking them, pulling them out of the ground? It's like, let's give them a little space." As a result she says she has several kinds of wildflowers not commonly seen in Connecticut these days. "If you turn the soil and see what comes up, you do get to see things that used to be more common."

I ask if any of the weeds in her backyard are invasive species. She pauses and smiles. "I do have a patch of *Polygonum cespitosum*. I think I probably put it there. I mean I think I probably introduced it on my clothing. So I feel like it's really not fair to get rid of it," she says. "But I don't let it spread as much as it would. I artificially maintain a balance, it's true. That is the human approach, right? Can never leave anything alone."

Chapter 11

———

Plant Futures

There are limits to saying,
In language, what the tree did.

—ROBERT HASS,
"THE PROBLEM OF DESCRIBING TREES," 2015

From the viewpoint of the evolutionary biologist it is
reasonable to assume that the sensitive, embodied actions of plants and bacteria are part of the same continuum of perception and action that culminates in our
most revered mental attributes. "Mind" may be the result of interacting cells. Mind and body, perceiving and
living, are equally self-referring, self-reflexive processes
already present in the earliest bacteria.

—LYNN MARGULIS AND DORION SAGAN,
WHAT IS LIFE?, 1995

Tony Trewavas has come nearly to the end of his career, and the future, in his estimation, is not bright. I've arrived at his 1800s farmhouse
outside Edinburgh because I want to hear from the scientist who has
been contemplating the nature of plants longer than any other, who

might be able to offer an answer to my lingering question of how we should think about them. It is late September, and the cab ride from the city wound past dry hills, barren except for a close shave of pale green grass, as if they'd all been given crew cuts. Trewavas's driveway is an oasis by comparison, with thick flowering bushes still in bloom despite the arrival of the fall chill, grown nearly up to the slate roof of the low-slung stone house. The visibility out of at least one full window has been ceded to the bushes entirely.

By this time I've met many researchers who studied specific parts of plants, or very specific things plants can do. Theirs were crucial but small peepholes into vegetal life. Trewavas, from everything I've read, prefers the wider view. He spends more time thinking of plants as whole beings, greater than the sum of their parts. Perhaps, I thought, he's already figured out what place plants should occupy in our minds and what it should change about how we live in the world. Trewavas is eighty-three years old and has been working as a plant biologist for sixty-four of those years, surely among the most of anyone still alive. He retired twenty years ago but is still churning out books and papers. He's contributed major discoveries in the world of plant hormones and signaling, and is now one of the foremost advocates for a rigorous treatment of the concept of plant intelligence.

But what I hear when we begin to talk is that he has become a committed pessimist about people. Trewavas has already apologized to his adult son for the world his generation and those previous have left him. Humanity has proven its failure as an evolutionary project and chosen a path of generalized destruction. Plants are intelligent, he has written over and over in books and papers. But we have been too slow to notice. It's probably too late now for that knowledge to change anything in the culture.

Hearing this in his living room on a dreary September day makes me want to close the book on our conversation altogether. I refuse to believe it is too late for anything, least of all for the world to awaken to the wonders of plants. I have; and it hasn't taken that long, in the scheme of things. But I've just arrived, so I don't leave; I stay and drink

the coffee his wife, Valerie, has given us on a tray full of cookies and
sweets. Both Trewavas and Valerie are wearing shades of blue. They
talk about the blue poppies Trewavas loves and plants along the drive
each year. They're blue like you wouldn't believe, Valerie says. "Like a
slice of the sky." I begin to see that Trewavas is not a complete pessimist,
at least not about all taxa. Plants, and their beauty, are something he
can still summon tremendous passion for. He speaks reverently about
a trip to California, where he saw the giant sequoias. "Awe, amazement,
unbelievable respect. You just stand and look, unable to take it in," he
says. "Like many I find touching such a monster is quite extraordinary."

Trewavas became a botanist because he couldn't stand killing rats
to do biochemistry research. Back when he was starting out, scientists
were expected to kill their own lab animals, often with the blunt force
of a metal ruler. It was how Valerie and Trewavas met; she was his student, and he offered to kill her rat for her, after her previous specimen
had bitten her on the hand. "My knight in shining armor," Valerie says.
But he always hated it. "Some people don't mind it, but I can't understand them. This is why I study plants," Trewavas says. I've now heard a
version of this from botanist after botanist; they study plants because
they couldn't bear the dark underbelly of animal work: most of the
time, you have to kill your subjects.

But of course, now, he's saddled with something else: respect for
plants. He's been publishing arguments for plant intelligence for decades now, despite plenty of criticism from more conservative peers.
But after decades of scrutinizing plants at an agonizing level of
minutiae—plant hormones are unbelievably complex—Trewavas became convinced that no amount of close attention to a single aspect of
a plant's physiology could tell the full story about what a plant is. In the
1970s, he came across *General System Theory*, a slim book by Ludwig
von Bertalanffy, which outlined the idea that biology was in fact an ag-
glomeration of systems or networks, which were all interconnected. He
shows me his worn copy, still on the shelf nearest his desk. This was the
dawn of network theory. The properties of organisms and populations
emerged from these connections, von Bertalanffy wrote—from many

parts interacting as a sort of whole. A plant is an emergent system like that, Trewavas decided back then. They're networks. This was heretical at the time, when biologists were focused on mechanistic discoveries about isolated plant parts. Thinking of plants as whole organisms led him to decide that plants are probably intelligent, and that intelligence is probably a property of all living things.* A brain is only one way to build a network, after all.

I ask him about his earlier assertion, that it's too late to change humanity's course, that it would require in part such a vast wholesale shift in the way people see plants as to be impossible. But what if it were possible? I feel compelled to ask. What would change? "I don't know what would happen if we managed to change people's perspective on plants," he says, looking thoughtful. I'm surprised he hasn't considered the ethical implications of that shift before, after all these years. I suppose that's the nature of pessimism; it forecloses the imagination of hope.

"Well, I would hope at least it would stop the hacking down of the rain forests. That is so short-sighted," Valerie says.

"Yes, the lungs of the planet, they call them," Trewavas says, sounding stricken. "I don't know why people are like this. It's about respect. If we'd respect plants more..." Tony trails off. "We find it not easy to feel the system in which we actually live," he says finally.

We can feel it a little, I think, even though we may not articulate it. It

---

* Tony: "Everything is intelligent. When people say they can't see it, they're talking about academic intelligence. They assume what they've heard in school about IQ and human intelligence is what it is. They ran with that for yonks ["For a long time," in UK slang]. That academic achievement, that's not about survival. What I'm talking about is not academic intelligence. It's biological intelligence. It doesn't seem to sink in, however many times I say it. It's silly, because it's not unique to plants. Every organism on this earth acts intelligently. When a zebra runs away from a lion, is that intelligent behavior? Of course it is, it's survival! And they don't have difficulty recognizing that. But when an insect bites a leaf and the plant produces a natural pesticide to ward it off, is that intelligence? It's no different. It's not running away from threats, but it's finding a way of survival. They don't connect the two."

can be as simple as the sense that there is something sacrilegious about felling a four-hundred-year-old tree for decking—or even a thirty-year-old pine for toilet paper. What did it take for that tree to live through those years, make thousands of leaves each spring, store sugars through the winter, turn light and water into layers and layers of wood? It is hard to underestimate the drama of being a tree, or any plant. Every one is an unimaginable feat of luck and ingenuity. Once you know that, you can't unknow it. A new moral pocket has opened in your mind.

The conversation turns to the reasons certain scientists so strongly reject the idea of plant intelligence. It's silly, Tony says. "In truth, scientists don't know enough about plants to make any dogmatic statement about them." We think they always photosynthesize, he says, but then we find parasitic plants that act more like mushrooms, and don't photosynthesize at all. Even the most basic of statements can be quicksand. There are no foregone conclusions. Except perhaps that evolution will find a way to flout any we come up with.

Still, despite the fact that Trewavas knows and publishes so much on the topic of plant intelligence, it surprises me that he hasn't thought much about what would happen in wider society if they were indeed accepted as such. It occurred to me that perhaps plant scientists are not the right people to come to for plant ethics. Philosophy and science have been separate skill sets for too long for that.

At the end of the day, whether or not plants are intelligent is a social question, not a scientific one. Science will continue to find that plants are doing more than we'd imagined. But then the rest of us will have to look at the data and come to our own conclusions. How will we interpret this new knowledge? How will we fit it into our beliefs about life on earth? That is the exciting part. Perhaps we'll decide it no longer makes sense to hold so tightly to our old beliefs about what plants are, given all this new information about their nature. Perhaps we will see them as the animate creatures they are.

But what happens then? Underlying all this is the deeper question, the one that matters most: What will we do with this new understanding? There are two directions to go in: we do nothing at all, and carry on as before, or we change our relationship with plants. At what point do plants enter the gates of our regard? When are they allowed in to the realm of our ethical consideration? Is it when they have language? When they have family structures? When they make allies and enemies, have preferences, plan ahead? When we find they can remember? They seem, indeed, to have all these characteristics. It's now our choice whether we let that reality in. To let plants in.

AFTER YEARS OF visiting plant scientists and reading about botany, my most luscious thoughts have all turned green. Plants have thoroughly gotten to me—but of course the reality is, they've had me all along. Plants made me, after all. Every bundle of muscle in my body was woven from the sugars plants spun from moisture and air. My blood cells that course through my veins like water through rootlets are each kept ruby red with the oxygen plants made. The branching structure of my lungs are suffused with that too. Every inward breath of mine was first breathed out by plants. In this material sense, in terms of what they've contributed to my physical being, they are as much my relatives as any family member I know.

Now when I spot a tendril that is making its way through a crack in the sidewalk, I internally commend it for its resourcefulness. I feel I know about some of what it took for the plant to do that—the small miracle of its germination, the craning of its elongation, the articulation of the hundreds, maybe thousands of fine root hairs, right now probing its belowground world for sustenance. I think about the stem cells in each of its growing tips, poised and ready to become whatever sort of flesh the plant needs them to be. The whole being a sensitive, decisioning network spread throughout hundreds of limbs, thousands of roots. A body in motion, adapting in real time to every subtle shift, flowing like water through its surroundings and taking note of the shape and smell and texture of it all.

It's a small gesture, my quiet acknowledgment, but I see it as a sign that something in my life has changed. I have come to view plants as creatures. I've brought them into the fold of animacy in my mind.

The hard evidence of plants' primacy is not hard to find, in a practical sense. The harder thing is to feel it. To begin to include plants in our vision of the moving, living world, and see them as animate individuals in their own right, takes mental effort. We may sense it, but many of us have been given neither the way of seeing nor the words to turn that feeling into fact.

One strain of philosophy argues that we should consider plants and other organisms to be conscious, and that our inability to do so is a willful lack of imagination. What would it be like, they wonder, if all organisms had a place in our society? The philosopher Bruno Latour once wrote that "in order to enroll animals, plants, proteins in the emerging collective, one must first endow them with the social characteristics necessary for their integration." But those "social characteristics" might not need to be endowed as such. Plants know their kin, they cooperate and fight, they mediate their relationships with one another and the other creatures that frame their lives. Perhaps it isn't just a philosophical exercise. Perhaps the social characteristics are already there. It seems to me now that they are.

Others have imagined the terrain that could exist beyond this mental hedge. In the world of Ursula K. Le Guin's 1974 short story *The Author of the Acacia Seeds*, it might be the year 2200, perhaps 2300. A major leap in human knowledge has taken place: animals of all types are found to have language, and not only language, but literature—art. A new field of linguistics has sprung up to translate them. Through careful study, therolinguists have uncovered the tunnel-sagas of earthworms, murder mysteries written in Weasel, and the "group kinetic texts" composed by pods of cetaceans as they move through underwater choreography. Some dialects, like the seed-arrangement-based language of Ant, can be translated directly into human text. Staging a group ballet seems to be the best way to translate the ineffable meaning of Adélie Penguin. Thousands of fish literatures are now known to humanity, and frogs seem particularly fond of writing erotica. These languages were always there, of course, but a crucial shift has taken place. Humans have learned how to see them.

But the president of the therolinguistics association wishes to draw attention to a massive oversight. Why has no therolinguist yet attempted to translate Plant? What might the redwood or zucchini be saying? New tools will be needed, as plants are likely to have entirely different orientations to the world. "But we should not despair," the president writes, in an editorial to their profession. "Remember that so late as the mid-twentieth century, most scientists, and many artists, did not believe that Dolphin would ever be comprehensible to the human brain—or worth comprehending!" The president imagines a group of future linguists laughing likewise at the present disinterest in the language of an eggplant "as they pick up their rucksacks and hike up to read the newly deciphered lyrics of the lichen on the north face of Pike's Peak."

It intrigues me that Le Guin's story came out just shy of a decade before David Rhoades published his discovery of the chemical conversation amongst red alders and Sitka willows in Washington.* We now know that plants do speak, in chemicals. Their health status, their assessment of risk in real time, and even the quality of their nectar is decipherable to us now by sampling the volatile chemicals they exude. They communicate with one another, and with members of other species when the situation calls for it. At what point do we decide that plant communication qualifies as language? What would it do to our own minds if we decide it does?

Perhaps plants are also speaking in movement, in electricity, or even in the flow of fluid in their bodies that clearly produces audible clicks, though all of this is yet to be understood. I think about Lilach Hadany putting her microphones up to grapes and wheat. We know animals

---

* A half-century after Le Guin wrote this story, scientists are now widely considered to be on the verge of translating whale language. Science fiction like Le Guin's has always been a tool for exploring otherness, inverting hierarchies of power, and questioning what we think we know. Plants are the ultimate other. As such they've long occupied a special place in sci-fi. For further reading, see *Plants in Science Fiction: Speculative Vegetation*, edited by Katherine E. Bishop, David Higgins, and Jerry Määttä (Cardiff: University of Wales Press, 2020).

can communicate through skin pattern changes, body waggles, hair ruffles, gestures. As soon as we reorient away from human modes of expression, we open ourselves to other worlds of being. We're learning more every year. Language may already be there, for a plant. We may not yet know how to hear it.

Science may never come fully to the conclusion that plants are intelligent, at least not in the way the word most readily, to our ears, implies. I'm beginning to wonder at what point that ceases to matter, given what we now know about them. *Intelligence* is a loaded word, perhaps overly connected to our ideas of academic achievement. It's been weaponized against fellow humans for millennia, used to divide people into hierarchies of worth and power. I wouldn't want to apply that schema to a whole additional category of life. Yet it is, by its very definition, still a word that contains the germ of what we mean by alert, awake to the world, spontaneous, responsive, decision-making. From the Latin *interlegere*: to discern, to choose between.

So science may or may not ever deign to use it for plants, for exactly the reasons of the social implications; humans have contaminated the word with their humanness. But words are merely symbols. They draw a perimeter around a feeling for which there is no language. In that sense, *intelligent* might be the tightest word-perimeter we've got to describe what we are seeing plants do. We can choose to nudge it back to its more universal meaning, its earlier Latin. But if not using the word is a social decision, made chiefly by skittish scientists hoping to cause no harm, its opposite can be too. We can take the risk, and hope that understanding will follow. We can try our best to make its meaning clear, not muddled by overly human categories. Putting too human a sheen on plant intelligence is a failure of imagination, after all. Plants are exuberantly, bafflingly, intelligently themselves.

This question of what words to use comes up so often that I've grown nearly tired of it.* At the center of the debate is the question of anthro-

---

* Consciousness is another, related matter. Consciousness can't be described or

pomorphizing; using human terms to describe plant lives. Some, like ecologist Carl Safina, argue that it's the "best first guess" we have as to what a nonhuman is experiencing. It lures the senses toward occupying other perspectives, a sort of bridge to understanding nonhuman lives. It's what the Greek philosopher Theophrastus, who coined the term *heartwood* to describe the inner flesh of trees, plainly advocated for: "It is by the help of the better known that we must pursue the unknown."

Doing anything else quickly becomes ridiculous. In a 2015 paper, anthropologist Natasha Myers noted that botanists were so wary of avoiding any whiff of anthropomorphic language that they resorted to ridiculous formulations to describe plants' lives. Instead of the plant "storing" starch and "mobilizing sugars" throughout the night, they wrote about how "the time of day of starch degradation is altered." A plant doesn't "react," instead it "is affected." The cardinal grammatical sin of passive voice is all over these botany papers. And it sounds absolutely terrible. Articulating these processes without ascribing agency

---

observed in the lab. This all comes down to the definitions we chose to use, all inventions of language that for all its admirable trying can't possibly encompass all that it means to be the alert, subjective beings we know ourselves to be. So if consciousness is the awareness of self, plants have it. So do single cells. If consciousness is the ability to be knocked unconscious, then it looks from the outside that plants have that too. Is there ever intelligence without some form of consciousness? My instinct says no. Perhaps we can split consciousness into fractions and degrees, with certain beings having more or less. A spectrum of consciousness. But then the word becomes unsatisfying, calls for other words. Words, at the end of the day, fail to capture biological creativity.

We are at a strange time in the history of understanding consciousness. Chatbots are starting to sound quite human. We are building intelligent machines which we can interact with as though they have consciousness. The question of the consciousness of these inanimate programs is all over the news. If we decide AI is in some way conscious, we will have made an inanimate thing animate. It implies that a mind can be coded for. This is all rather disenchanting. It suggests a predetermined, unwilful world of the mind, and does nothing to explain the subjectivity we feel inside ourselves, whether or not it can be measured or explained.

is actually quite difficult, fumbly, imprecise.* When Myers asked a researcher whether she thought plant structures could be considered analogous to the human nervous system, the researcher said no. She thinks trying to use human language for plants "cheapens plants," because it "assumes we are the ultimate being." Instead, plants are far more advanced than humans in multiple categories. Take the remarkable fact that they can produce complex chemicals, like caffeine. "These are skills we don't have," the researcher says. Comparing them to humans erases those capacities.

But I wonder, instead of humanizing plants, could we not just vegetalize our language? We can call these traits plant-memory, plant-language, plant-feeling. The plant-specific essence of each word would stand behind it like a ghost. If plants are intelligent in their own, vegetal manner, perhaps we call that plant intelligence. It rolls right off the tongue.

It is most clear to me that including plants in our ethical imaginations will have to be a social choice when I think about the very recent history in which performing surgery demonstrations on living, unanesthetized dogs was the norm. Doctors and scientists justified this because animals, they believed, couldn't experience pain. This idea sounds plainly ridiculous and abjectly cruel to us now, but science said otherwise back then. The ultimate turn away from vivisection began not because the surgical profession changed its mind, but because social tides, led by the first humane societies, had turned against the practice.†

---

* I noticed this too. Time and time again, papers use passive language to talk about what a plant does. But when I visited scientists in their labs or fields, they gamely anthropomorphized the plants, talking among themselves about how a plant "hates that" or noting that a certain treatment "makes them happy." I knew these scientists didn't picture their plants as little humans when they said that. They of all people most intimately knew that plants were their own thing entirely. They'd just already resolved the discrepancies in their minds, and widened the language to suit this other category of being.

† Organizations of mostly women founded the first humane societies, advocating for the rights of animals. These women appealed to the hearts and minds of enough people that vivisection became socially unacceptable. Many of them became suf-

To some, leaping to the ethical consideration of plants when animal rights are hardly secured is a preposterous distraction. After personally being berated by friends and colleagues in animal studies for suggesting that plants might be ethically compelling, too, Jeffrey T. Nealon wonders if this "seems to function as a subset of an old practice: trying to close the barn door of ethical consideration right after your chosen group has gotten out of the cold of historical neglect." It's a story that gets repeated over and over. But moral attention is not a finite resource.

This business of drawing a line between what does and doesn't deserve our respect and attention can feel like an exercise in absurdism. It produces a great deal of cognitive dissonance in me now. So what if, I wonder, plants had a place in our society? What would an ethics that includes plants look like?

One place to begin to contemplate this is in law. In 1969, the Sierra Club sued to block the Walt Disney Company from embarking on plans to build a ski resort in an area of a subalpine glacial valley abutting Sequoia National Park. The resort would cost double what the original Disneyland had cost to build, and would require the construction of a twenty-mile highway to lead fourteen thousand visitors into the valley each day. The suit made it to the Supreme Court, but the court rejected it in 1972 on the basis that the Sierra Club did not have standing: they would not personally be injured by the resort in any way. In his dissent, Justice William O. Douglas memorably writes that plants and ecological entities ought to be able to sue in matters of their own protection:

> Inanimate objects are sometimes parties in litigation. A ship has a legal personality, a fiction found useful for maritime purposes. . . . So it should be as respects valleys, alpine meadows,

---

fragettes, advocating for the right of women to vote, another idea considered institutionally ridiculous until suffragettes forced a social shift in what was considered acceptable. Indeed, the end of vivisection and the beginning of votes for women are connected: as the circle of rights widens, it's hard to see why it shouldn't keep widening.

rivers, lakes, estuaries, beaches, ridges, groves of trees, swamp-land, or even air that feels the destructive pressures of modern technology and modern life.... The voice of the inanimate object, therefore, should not be stilled.

That same year, in an essay titled "Should Trees Have Standing?" legal scholar Christopher Stone ruminates on the idea that legal rights for plants may seem, at present, "unthinkable." But, he writes, humans have been in the business of expanding legal rights to new groups forever. Often this comes after long periods of excluding those same entities from rights by arguing that their exclusion is only "natural." In the United States, legal rights for groups of people, like Black people, Chinese people, Jewish people, and women, were considered "unthinkable" by many at the time those rights were granted. Nonhuman entities like corporations, trusts, nation-states, and even ships—"Still referred to by courts in the feminine gender"—have had legal standing in some cases much longer than some of those groups of people. Still, jurists who watched corporations gain rights in the courts argued that this was "unthinkable" too. And if corporations can be assigned rights, Stone argues, trees ought to be given them too. Unthinkability is hardly an excuse.

"I am quite seriously proposing that we give legal rights to forests, oceans, rivers and other so-called 'natural objects' in the environment—indeed, to the natural environment as a whole," Stone writes; at different points in history, our social "facts," on which law is often based, have changed. We create a collective "myth" of ourselves and the world, he writes, which reflects our present norms, and is enshrined in our laws. But we tend to forget that these norms are fabrications. "We are inclined to suppose the rightlessness of rightless 'things' to be a decree of Nature, not a legal convention acting in support of some status quo," Stone writes. As our knowledge of "geophysics, biology and the cosmos" grows, so should our collective "myth" and our laws expand with it. The status quo as such is outmoded. It's time for something new. I find myself wondering what Stone would think of the recent advances in botany, which have uncovered so much about plants that was unthinkable hardly two decades ago. I'm sure he would all the more vigorously agree that plants deserve legal personhood. In fact, it's overdue.

It was with this in mind that I watched from afar as wild rice sued the state of Minnesota. The 2021 suit was brought by an attorney for the White Earth Band of Ojibwe who represented the wild rice that grows in the shores of northern Minnesota's wetlands, and which was being threatened by a pipeline slated to barrel right through its habitat. The state had given Enbridge, a Canadian company, a permit to construct its pipeline without consulting the White Earth nation, who held treaty rights to harvest the wild rice in the area. The grain is central to the lives of the Ojibwe, and every September, ricers canoe through the shallows to harvest it.

Wild rice requires an abundance of very clean water in order to grow. The pipeline would bring tar sands crude from Canada right through wild rice habitat, and with it, the threat of a spill. So the White Earth Band of Ojibwe gave the rice legal standing, recognizing its "inherent rights to exist, flourish, regenerate, and evolve." The right to evolve! I'd never seen such biologically expansive language in a lawsuit. It looked like a historic moment for plants as legal persons. But the tribe's own court dismissed the case in 2022 for lack of legal precedent.

Legal personhood for plants may still be around the corner yet.* But plant personhood itself is a concept as old as human culture. As we've already learned, Native philosophies from all corners of the globe often understand plants as relatives, or ancestors, or otherwise persons in their own right. It's not that plants are human, but that humans are just one kind of person, as are animals. Personhood means one has agency and volition, and the right to exist for their own sake. Harming an ani-

---

* Meanwhile, the Kichiwa of Sarayeku, an Indigenous group in Ecuador, are currently advocating for the United Nations to recognize their forest territory in the Amazon rain forest as a conscious being endowed with universal rights.

mal person (or plant person) may be crucial to one's ability to survive, but it can't possibly be disregarded. Yes, you have to eat. You have to make clothing and build houses. You have to kill plant persons and animal persons to do that. That's a fact of life. But that doesn't leave any excuse for indiscriminate killing, or thoughtless destruction.

Plants in Indigenous philosophy and cosmology are often literal relatives and ancestors. Indigenous Maya people in present-day Mexico believe the first people were made from corn. In virtually all cosmologies, plants and people are descended from the same extended ecological ancestry. This is, of course, now understood as an evolutionary fact. We certainly do share a common ancestor with plants, though it was a long time ago. What would it be like for that fact to feel less remote, and more present in one's life? It links everyone in a certain relationality, an extended kinship. If a plant is a person, they are entitled to their autonomy. An encounter with a plant is an encounter between two beings. Deborah Bird Rose called this an "intersubjective encounter." When you think of a plant like that, a profound moral force enters the space between you, sticking to everything like spider silk. You can't ignore it, you can't step outside of it. One might call it respect.

Respect comes with a certain responsibility of care, of maintaining a good relationship. Plant personhood may be something we have to teach ourselves, and perhaps we initially strain to see it. But once seen, the care part of that new awareness comes quite naturally. You might find you respect the autonomy of plants not because you know you "should," but because you know you must. Because doing otherwise would violate your own moral personhood. It is a bridge to cross, from plant disregard to plant regard. The distance between the two is the orientation of one's heart on the subject.

Of course, we've come such a long way to arrive at much the same place as so many people already have. But new findings in botany are opening up a chance to remodel the way we see the nonhuman world, and our place in it. It occurred to me that all this hand-wringing among scientists about what to call plants is a faithlessness in the public's imagination. That they'd take it too far, absorb too simple a message,

begin to see plants as cartoon characters, or the equivalent of tiny om-
niscient demi-gods. That's not an unreasonable fear; I get it, sometimes
the simplest possible message is the one that gets through. But as a
journalist, I feel acutely aware of the danger of eliminating nuance and
complexity for fear that the message won't be digestible.

It's that lack of faith in the public that always results in an erosion
of the level of public discourse. A faithlessness in the public is a self-
fulfilling prophecy. Remove complexity, and the capacity for complexity degrades farther. I think people can be trusted to handle a
complicated truth. Plants are not omnipotent, otherworldly creatures.
They are also not just like us. But neither are they neither of these
things. There are elements of reality in both images, and fallacy in both
too. This is hard stuff: one needs to welcome ambiguity and delight in
the lack of easy tropes. Complexity is the rule in nature, after all. Thinking through this requires occupying a mental space of in-betweenness
rarely tolerated in our contemporary world concerned with linear narratives and known entities.

Báyò Akómoláfé, a Yoruba poet and philosopher, wrote about this
in-betweenness, contemplating the way all creatures are in fact composite organisms. The state of nature is one of interpenetration and
mingling that defies easy categorization. It occupies a middle place,
both in the material reality of the world and in our understanding of
it. "The middle I speak of is not halfway between two poles; it is porousness that mocks the very idea of separation," he writes. Akómoláfé
outlines our collective biological reality as a state of "brilliant betweenness" that "defeats everything, corrodes every boundary, spills through
marked territory, and crosses out every confident line." It reminds me
of Trewavas, telling me in his living room outside Edinburgh that
scientists don't know enough about plants to say anything dogmatic
about them. Scientists know a tremendous amount about plants. They
also might not yet know what a plant is at all.

Akómoláfé's description of brilliant betweenness applies to everything I'd come to understand about plants, and about us. Our idea of
them needs to exist in the shimmering, porous place in our mind. It's a

tough place to access. Perhaps you haven't used yours since childhood. It's hard to exist in that middling place. But it's not impossible. I've stepped beyond that gate. I have faith that others can too.

Some will consider this a matter of philosophy and belief, which they might say always signifies a departure from science. But it's not that science is being drawn ever farther from itself. Rather, the space between science and ethical meaning is being stitched together with webbing. Thin tendrils are amassing a delicate bridge.

The miraculous thing about this is that the entire world could be different. Seeing plants as beings worthy of rights would open up a different realm of being with plants. It would revolutionize our moral system, our legal system, and the way we live on this earth.

SOMETIME AFTER MY trip to Scotland, I found myself deep inside a cave in Puerto Rico. The interior of the island is mountainous, covered in lush jungle so dense that only green light filters through to the forest floor. But this jungle has many mouths, like stomata on the back of a leaf, if you know where to look. The rocky mouths open into darkness: entrances to a vast cave system that threads beneath the island in a vasculature of underground rivers and rooms.

Luckily, friends here did know where to look. Ramón and Omar found the particular mouth they wanted, and we began descending into its cold blackness, leaving the midday warmth behind. We passed ancient Taino petroglyphs carved into stalactites—stacks of round faces and lizards and spirals—where the last bit of green light still penetrated, and then suddenly it was totally dark. Headlamps were turned on. As we got deeper, roots from above followed us. In one cavernous room, taproots as thick as my forearm sent by trees on the surface had drilled themselves through feet of solid rock above our heads to emerge into this black cathedral, spooling another thirty or forty feet through empty air to reach their target: the gently flowing underground river we edged alongside. Such tremendous effort for a drink. It seemed excessive. But I trusted there was some logic to this hefty infiltration that my human eyes just couldn't understand.

We traveled away from the river. We'd been walking underground for three or four hours now, sometimes slithering on our bellies to fit through openings in the rock no wider than our hips. But the deprivation of complete darkness expands time in strange ways, and I felt like I'd been there forever, like I might never see daylight again. This was not a realm for people, though I knew people had been occasional guests here for millennia, the modern history of which was scrawled in graffiti along the cave walls, signatures from 1914 and 1939 and 1974. Strange insects clung to the walls too, including one with big black claws, which, Ramón told me, carries its eggs in little pockets on its back. When the babies hatch, they burst through their parent's back, shredding it to pieces. Once, the sweep of my headlamp caught sight of a scorpion. I chose not to think about it.

We crawled into the next room and stood up. Immediately my sneakers sunk into a loamy mousse. The air smelled musty and faintly sweet. Something squeaked. I looked up. Hundreds of fruit bats hung upside down from pocks in the ceiling, quivering fitfully against each other like bristles on an agitated hedgehog, but round and fluffy and very cute. One stretched out its wing and retracted it again. The skin was so thin it let some of the light from my headlamp through. I finally understood the mousse. We were standing in bat guano.

I looked around at the blanket of bat excrement. Hundreds of white chopsticks appeared to be growing out of it, directly at the center of the room, on the patch of ground right beneath the biggest quiver of bats. The slender stems, a foot tall and pure white, were each topped with a single white leaf, or sometimes two, like a flag on a toy sailboat. I realized this was the bats' doing: these fruit-eaters returned to their cave after a night feasting in the forest above and excreted seeds, probably by the thousands. Fruit bats are some of the most important seed dispersers in this ecosystem. But that only works out the way the plants intended when the bats drop their seeds aboveground. Down here, the plants were doomed. There was no light, so no photosynthesis. No chance of life-giving green. There was only the bewitching fertility of the most potent fertilizer on earth; a foot-deep bed of bat guano.

This was a ghost forest, haunting in its inevitable futility. The fuel from their seeds would run out, they would die soon. Why did they bother coming up at all? I wondered. After everything I'd learned about the extreme good sense of plants, it seemed to me that I was looking at an example of plant stupidity.

Yet it was also somehow relatable. I looked again. It was clear they'd tried their absolute best. They'd grown as tall and as slender as structurally possible, using all of their finite energy to look for any scrap of light. They only put out one or two leaves—pennants of hope that some photon might fall upon them after all. Their strategy was remarkably sensible. I didn't know if all the seeds in this cluster were from the same species of plant or not; fruit bats tend to eat lots of different fruits, so it seemed unlikely. Yet every single one of these white seedlings looked the same. Perhaps they'd all converged on the same form because this was the best one for survival. They adapted as best they could to their situation, and put everything they had into the wisest possible shape.

It still wasn't enough, but that wasn't the point. Perhaps intelligence of any kind is not measured in success, but in the approach. Would any of us, were we the plants in this situation, done any differently? They tried to survive in an inhospitable landscape the best way they knew how. I felt moved by this. This was a craning toward life, even in the face of impossible conditions.

Our own humanity is found as much in our achievements in a harsh and complex world as it is in our limitations, our frailties, our faults. We are no less human for them. Perhaps this amorphous quality of "plant intelligence" I'm trying to grasp—this liveliness and beinghood plants unequivocally have—is as much to do with plant trying, plant testing, plant failing. After all, who we are shows itself not just in the outcomes of our goals but in the paths we take to get there. The trying says more about what is inside us than does the success.

Again, there are no foregone conclusions. If I've learned anything, it's that biotic creativity is our inheritance. Rather than seeing a march toward doom, as I did as a disaffected office worker writing the news, I now see a boundless sea of change. Life finds a way, if given a chance.

But what happens when that chance is given, or removed, by us? The wellbeing of plant communities globally now depends on human attitudes toward them. Now that we can see plants as individuals, we have learned to see them on their own terms. Perhaps now we can fold that special admiration back into the larger whole. Biologically, their value lies in their function as members of interrelated communities, the rich, interspecific interactions that hold up the world, the one in which we are all a part.

A single plant is a marvel. A community of plants is life itself. It is the evolutionary past and future entangled into a riotous present in which we are ourselves also entangled. This stretches the mind. Plants give us the chance to see the system in which we live.

Acknowledgments

One winter afternoon in 2018, I was tucked into a corner booth in a pub on the west coast of Ireland with Sarah Grose, my oldest friend. It was half past four and already dark outside. I had the feeling of being on the edge of something new. I told her I thought I wanted to write a book. Something about plants. It was the first time I'd said that aloud. Sarah told me to write that thought down, right now in this pub, because, she said, you really are going to do that. Thank you for always knowing me before I know myself.

In the years that followed, so many people came into my life to shape this book. A book is a solitary pursuit only in the purely mechanical sense. Many dozens of scientists gave me hours of their time, some across years, and several welcomed me into the places where they worked. In these exchanges I was always aware of the humbling fact that everything a scientist knows is the product of untold hours in the lab and decades inside the gauntlet of academia. And the ones in this book did all that in the name of plants. Imagine. My deepest thanks to each of you for your generosity with me. In particular I want to thank Rick Karban, Liz Van Volkenburgh, Ernesto Gianoli, and JC Cahill. Our long-running correspondence has been truly invaluable.

To Adam Eaglin, my agent, who lifted this book up from the start: as a debut author I never imagined this level of professional support was possible, let alone given with such total elegance as you offer it. You are a writer's hero. I and all your clients are so lucky to have you. Thank you also to the entire team at the Cheney Agency, including the great Elyse Cheney herself, for such thorough and unwavering support.

I'm profoundly grateful to Sarah Haugen, my editor at Harper, whose questions and critiques elevated this book immeasurably. Your notes of encouragement buoyed me through every draft. Thank you for really getting it. And to Gail Winston, who first believed in this project: your early wisdom and deep sense of craft made me feel in the best of hands. Thank you for first making me feel like an author. Thank you to Milan Bozic for this perfectly alien cover design, Maya Baran for her publicity prowess, and the rest of the great team at Harper, who have been such incredible champions since day one. To Emily Krieger, my excellent fact-checker: I'm so lucky to be checked by a fellow plant person.

My eternal gratitude to the artist residencies, that gave me space to write in glorious locations, several of which made it into this book. Each of them taught me how to listen more closely to the landscape and myself and to take seriously what emerges. Thank you to The Mesa Refuge in Point Reyes, California, Bloedel Reserve on Bainbridge Island in Washington, The Strange Foundation in West Shokan, New York, The Marble House Project in Dorset, Vermont, The Folly Tree Arboretum in Easthampton, New York, Oak Spring Garden Foundation in Virginia, and the National Parks Arts Foundation for a month in Hawai'i Volcanoes National Park on the Big Island. Thank you in particular to the Park Service botanists and ecologists I met there; I learned so much from you. Thank you also to the National Tropical Botanical Garden on Kaua'i, whose environmental journalism fellowship first introduced me to Steve Perlman and sent me down a new path. To Lincoln, Cody, Laura, and Farmer Bill Hill: the months at your farm were some of the happiest of my life.

To Lucy McKeon, Julia Simpson, Nadja Spiegelman, and Carina del Valle Schorske, you are each brilliant in both creativity and friendship, and I'm so lucky to be the beneficiary of both. Thank you for your close reads, your writerly guidance, your sharp critiques, and everything I learn when I'm in your company.

Thank you to my friends Lily Consuelo Saporta Tagiuri, Jaffer Kolb, Ryan Moritz, Nikhil Sonnad, Althea SullyCole, Suzanne Pierre, Zoe Mendelson, Rose Eveleth, Olivia Gerber, Annabelle Maroney, Joseph Chugg, Olaya Barr, and many other friends both new and old for conversations that fertilized my thinking throughout the years of this project and long before it. I am a product of time spent with all of you.

Thank you to my mother, D, my biggest supporter. You know how to find the magic in everything the world has to offer. I get all my open-minded curiosity from you. To my father, Rafe, thank you for showing me that physics and biology could be astounding when I was very young; your sense of awe at the mechanics of the world clearly left its impression on me. To my brother, Mikolo, your gentle openness and inventiveness inspire me. I love you.

Thank you to Marleen DeGrande, my grade-school teacher, who taught me about poetry and persistence, and how to turn thinking into action. You once told me I should be an artist. I do hope this counts.

To Anne Humanfeld and Jeff Schlanger, to whom this book is dedicated: thank you for a lifetime of learning how best to love the world through reverent appreciation of its beauty. Your perspective on all things has profoundly shaped my own.

Most of all, thank you to Sarah Sax. Every author knows the rough emotional topography of a book project. I stubbornly believe none are as lucky as I was to have Sarah's curative optimism and complete care to return to. She has lived this project with me all the way through and fertilized it with her own curiosity and intellect. Our shared interest in the natural world is a renewing pleasure, a spring-fed well. So many of the ideas here first debuted in conversation with her. Many of her suggested readings expanded my brain and twined their way into these pages. Sarah, you are my first reader and favorite editor. Our life together is the most interesting extended conversation I've ever had the privilege to be party to. With you, all things feel possible, and something new is always just around the corner.

Notes

Chapter 1: The Question of Plant Consciousness

7   taste with their arms: Lena van Giesen et al., "Molecular Basis of Chemotactile Sensation in Octopus," *Cell* 183, no. 3 (2020): 594–604.

7   use tools: Jennifer Mather, "Cephalopod Tool Use," in *Encyclopedia of Evolutionary Psychological Science*, ed. Todd Shackelford and Vivian Weekes-Shackelford (New York: Springer, 2021): 948–51.

7   remember human faces: Roland C. Anderson et al., "Octopuses (Enteroctopus dofleini) Recognize Individual Humans," *Journal of Applied Animal Welfare Science* 13, no. 3 (2010): 261–72.

9   a complete fern genome had been sequenced: Fay-Wei Li et al., "Fern Genomes Elucidate Land Plant Evolution and Cyanobacterial Symbioses," *Nature Plants* 4, no. 7 (2018): 460–72.

9   720 pairs of chromosomes: D. Blaine Marchant et al., "Dynamic Genome Evolution in a Model Fern," *Nature Plants* 8, no. 9 (2022): 1038–51.

10   Farmers in China and Vietnam: Thomas A. Lumpkin and Donald L. Plucknett, "Azolla: Botany, Physiology, and Use as a Green Manure," *Economic Botany* 34 (1980): 111–53.

15   "corpus of fallacious or unprovable claims": Arthur W. Galston and Clifford L. Slayman, "The Not-So-Secret Life of Plants: In Which the Historical and Experimental Myths about Emotional Communication between Animal and Vegetable Are Put to Rest," *American Scientist* 67, no. 3 (1979): 337–44.

16   find themselves beside their siblings: María A. Crepy and Jorge J. Casal, "Photoreceptor Mediated Kin Recognition in Plants," *New Phytologist* 205, no. 1 (2015): 329–38.

16   able to hear water flowing: Monica Gagliano et al., "Tuned in: Plant Roots Use Sound to Locate Water," *Oecologia* 184, no. 1 (2017): 151–60.

16   lima beans: Junji Takabayashi, Marcel Dicke, and Maarten A. Posthumus, "Induction of Indirect Defence Against Spider-Mites in Uninfested Lima Bean Leaves," *Phytochemistry* 30, no. 5 (1991): 1459–62.

16   tobacco: Silke Allmann and Ian T. Baldwin, "Insects Betray Themselves in Nature to Predators by Rapid Isomerization of Green Leaf Volatiles," *Science* 329, no. 5995 (2010): 1075–78.

16   hungry caterpillars: John Orrock, Brian Connolly, and Anthony Kitchen, "Induced Defences in Plants Reduce Herbivory by Increasing Cannibalism," *Nature Ecology and Evolution* 1, no. 8 (2017): 1205–7.

17   Bernard Berenson, *Sketch for a Self-Portrait* (New York: Pantheon, 1949), 27.

18   "Plants Neither Possess nor Require Consciousness": Lincoln Taiz et al., "Plants Neither Possess nor Require Consciousness," *Trends in Plant Science* 24, no. 8 (2019): 677–87.

20   "This is surely circular reasoning": Paco Calvo and Anthony Trewavas, "Physiology and the (Neuro) Biology of Plant Behavior: A Farewell to Arms," *Trends in Plant Science* 25, no. 3 (2020): 214–16.

22 naturalists believed: Joseph Priestley, "Letter to Benjamin Franklin from Joseph Priestley, 1 July 1772," Founders Online, National Archives, https://founders .archives.gov/documents/Franklin/01-19-02-0136

Chapter 2: How Science Changes Its Mind

25 "Facts are theory-laden": Haraway, Donna J., "In the Beginning Was the Word: The Genesis of Biological Theory." *Signs* 6, no. 3 (1981): 469-81. http://www.jstor.org /stable/3173758.

25 "To ask humankind what being in the world means": Emanuele Coccia, *The Life of Plants: A Metaphysics of Mixture* (Hoboken, NJ: John Wiley, 2019).

26 The first plant was born a chimera: Frederick W. Spiegel, "Contemplating the First Plantae," *Science* 335, no. 6070 (2012): 809-10.

26 that first cyanobacteria is still within them: G. M. Cooper, *The Cell: A Molecular Approach*, 2nd ed. (Sunderland, MA: Sinauer, 2000). Chloroplasts and Other Plastids.

26 80 percent of Earth's living matter: Yinon M. Bar-On, Rob Phillips, and Ron Milo, "The Biomass Distribution on Earth," *Proceedings of the National Academy of Sciences* 115, no. 25 (2018): 6506-11.

26 By incessantly breathing out: Timothy M. Lenton et al., "Earliest Land Plants Created Modern Levels of Atmospheric Oxygen," *Proceedings of the National Academy of Sciences* 113, no. 35 (2016): 9704-9.

35 include plants in their family structures: Theresa L. Miller, *Plant Kin: A Multispecies Ethnography in Indigenous Brazil* (Austin: University of Texas Press, 2019).

36 *Plants Have So Much to Give Us*: Mary Siisip Geniusz, *Plants Have So Much to Give Us, All We Have to Do Is Ask: Anishinaabe Botanical Teachings* (Minneapolis: University of Minnesota Press, 2015), p. 21.

36 But *vegetabilis* came from the medieval Latin: Michael Marder, *Plant-Thinking, A Philosophy of Vegetal Life*. (New York: Columbia University Press, 2013).

36 "The philosophical project of naming": Jane Bennett, *Vibrant Matter: A Political Ecology of Things* (Durham, NC: Duke University Press, 2010).

37 to Plato, there was no sensation without intelligence: Amber D. Carpenter, "Embodied Intelligent (?) Souls: Plants in Plato's Timaeus," *Phronesis* 55, no. 4 (2010): 281-303.

37 The rational thing, then, was for men to rule: Women, children, and slaves had desiring souls, not reasoning souls, as elaborated by feminist academic Val Plumwood. Plumwood, *Feminism and the Mastery of Nature* (New York: Routledge, 1993), 84-85, quoted in Matthew Hall, *Plants as Persons: A Philosophical Botany* (Albany: SUNY Press, 2011).

37 Keeping the peace no longer required ritual acts of deference: Matthew Hall, "The Roots of Disregard: Exclusion and Inclusion in Classical Greek Philosophy," in *Plants as Persons*, 17-36.

37 first known texts about plants themselves: Theophrastus, *Historia plantarum*, c. 350-c. 287 BC.

38 weigh their shorter lifespans as a reasonable trade-off: Theophrastus, *De causis plantarum* 1.16.12.

38 "heartwood": Theophrastus, *Historia plantarum* 1.2.7-1.2.8.

38 "by the help of the better known": Theophrastus 1.2.5.

39 the notion of the "animal machine": Gary Hatfield, "Animal," in *The Cambridge Descartes Lexicon*, ed. Lawrence Nolan (Cambridge, U.K.: Cambridge University Press, 2015), 19-26, doi:10.1017/CBO9780511894695.010.

39 "vital phenomena": Thomas Huxley, "On the Hypothesis that Animals are Automata, and Its History," *Nature* 10 (1874): 362-66.

40 Ibn al-Nafis... beat him by a long shot: Mohd Akmal, M. Zulkifle, and A. H. Ansari, "Ibn Nafis-A Forgotten Genius in the Discovery of Pulmonary Blood Circulation," *Heart Views: The Official Journal of the Gulf Heart Association* 11, no. 1 (2010): 26.

40 discovering that bats navigate by echolocation: Carol Kaesuk Yoon, "Donald R. Griffin, 88, Dies; Argued Animals Can Think," *New York Times*, November 14, 2003.

41 cats exhibit the same attachment styles as human toddlers: Kristyn R. Vitale, Alexandra C. Behnke, and Monique A. R. Udell, "Attachment Bonds between Domestic Cats and Humans," *Current Biology* 29, no. 18 (2019): R864–R865.

41 scientists gathered . . . to formally confer consciousness: Philip Low et al., "The Cambridge Declaration on Consciousness," paper presented at Francis Crick Memorial Conference, Cambridge, England, 2012, 1–2.

41 Lizards have been shown to . . . navigate mazes: Lara D. LaDage et al., "Spatial Memory: Are Lizards Really Deficient?," *Biology Letters* 8, no. 6 (2012): 939–41.

41 Honeybees . . . distinguish between styles of art: Wen Wu et al., "Honeybees Can Discriminate between Monet and Picasso Paintings," *Journal of Comparative Physiology A* 199 (2013): 45–55.

41 "waggle dance": Shihao Dong et al., "Social Signal Learning of the Waggle Dance in Honey Bees," *Science* 379 (March 2023): 1015–18.

42 bees may have a form of subjectivity: James Gorman, "Do Honeybees Feel? Scientists Are Entertaining the Idea," *New York Times*, April 18, 2016.

46 resistance by scientists to scientific discovery is a known fact: Bernard Barber, "Resistance by Scientists to Scientific Discovery," *American Journal of Clinical Hypnosis* 5, no. 4 (1963): 326–35.

47 In a controversial article: Eric D. Brenner et al., "Plant Neurobiology: An Integrated View of Plant Signaling," *Trends in Plant Science* 11, no. 8 (2006): 413–19.

48 "New concepts are needed": František Baluška and Stefano Mancuso, "Plants and Animals: Convergent Evolution in Action?," in *Plant-Environment Interactions: From Sensory Plant Biology to Active Plant Behavior*, ed. František Baluška (Berlin: Springer, 2009), 285–301.

49 wrote Lincoln Taiz: Lincoln Taiz et al., "Plants Neither Possess nor Require Consciousness," *Trends in Plant Science* 24, no. 8 (2019): 677–687.

51 there may be no wizard behind the curtain: Michael Pollan, "The Intelligent Plant," *New Yorker*, December 15, 2013.

Chapter 3: The Communicating Plant

53 the paper that changed botany forever: David F. Rhoades, "Responses of Alder and Willow to Attack by Tent Caterpillars and Webworms: Evidence for Pheromonal Sensitivity of Willows," in *Plant Resistance to Insects*, ed. Paul A. Hedin (Washington, DC: American Chemical Society, 1983), 55–68.

55 "This suggests that the results may be due to airborne pheromonal substances!": David F. Rhoades, "Responses of Alder and Willow to Attack by Tent Caterpillars and Webworms: Evidence for Pheromonal Sensitivity of Willows," in *Plant Resistance to Insects*, ed. Paul A. Hedin (Washington, DC: American Chemical Society, 1983), 3.

55 information must be passed to it by its collaborator cells: Anthony Trewavas, *Plant Behaviour and Intelligence* (Oxford, U.K.: Oxford University Press, 2014), 48.

55 how a sperm and an egg self-organize to make us: Cells demonstrate such a "bewildering" range of responses to a "startling" variety of inputs that arguments have recently been made that they, too, should be said to be able to learn. See Sindy K. Y. Tang and Wallace F. Marshall, "Cell Learning," *Current Biology* 28, no. 20 (2018): R1180–84.

56 "knowledge the cell has of itself": From McClintock's Nobel address: "A goal for the future would be to determine the extent of knowledge the cell has of itself, and how it utilizes this knowledge in a 'thoughtful' manner when challenged." Barbara McClintock, "The Significance of Responses of the Genome to Challenge," *Cell Science* 226, no. 4676 (1984): 792–801.

56 "decision-making center": Alexander T. Topham et al., "Temperature Variability Is

Integrated by a Spatially Embedded Decision-Making Center to Break Dormancy in *Arabidopsis* Seeds." *Proceedings of the National Academy of Sciences* 114, no. 25 (2017): 6629–34.

56 agreed-upon definition for what counts as communication: Richard Karban, *Plant Sensing and Communication* (Chicago: University of Chicago Press, 2015).

59 bludgeoned by colleagues in journals: Simon V. Fowler and John H. Lawton, "Rapidly Induced Defenses and Talking Trees: The Devil's Advocate Position," *American Naturalist* 126, no. 2 (1985): 181–95.

59 paper by pioneering cicada researcher JoAnn White: J. White, "Flagging: Hosts Defences versus Oviposition Strategies in Periodical Cicadas (Magicicada spp., Cicadidae, Homoptera)," *Canadian Entomologist* 113, no. 8 (1981): 727–38.

60 placed pairs of sugar maple: Ian T. Baldwin and Jack C. Schultz, "Rapid Changes in Tree Leaf Chemistry Induced by Damage: Evidence for Communication between Plants," *Science* 221, no. 4607 (1983): 277–79.

60 could never quite get the trees to do it again: Peter Frick-Wright, "Early Bloom," interview with Jack Schultz, podcast, Public Radio Exchange, August 8, 2014.

61 pea-crossing studies in hawkweeds: Gian A. Nogler, "The Lesser-Known Mendel: His Experiments on *Hieracium*," *Genetics* 172, no. 1 (2006): 1–6.

65 goldenrods … issue chemical alarm calls: Aino Kalske et al., "Insect Herbivory Selects for Volatile-Mediated Plant-Plant Communication," *Current Biology* 29, no. 18 (2019): 3128–33.

66 Sagebrush use "private" means of communication: Patrick Grof-Tisza et al., "Risk of Herbivory Negatively Correlates with the Diversity of Volatile Emissions Involved in Plant Communication," *Proceedings of the Royal Society* B 288, no. 1961 (2021): 20211790.

66 extremely specific song phrases: Pamela M. Fallow and Robert D. Magrath, "Eavesdropping on Other Species: Mutual Interspecific Understanding of Urgency Information in Avian Alarm Calls," *Animal Behaviour* 79, no. 2 (2010): 411–17. Mylène Dutour, Jean-Paul Léna, and Thierry Lengagne, "Mobbing Calls: A Signal Transcending Species Boundaries," *Animal Behaviour* 131 (2017): 3–11.

67 science has begun to take seriously… animals have personalities: Jonas Stiegler et al., "Personality Drives Activity and Space Use in a Mammalian Herbivore," *Movement Ecology* 10, no. 1 (2022): 1–12.

70 male plants will only listen to male plants: Xoaquín Moreira et al., "Specificity of Plant-Plant Communication for *Baccharis salicifolia* Sexes but Not Genotypes," *Ecology* 99, no. 12 (2018): 2731–39.

70 sagebrush plants will listen to their genetic kin: Richard Karban et al., "Kin Recognition Affects Plant Communication and Defence," *Proceedings of the Royal Society* B 280, no. 1756 (2013): 20123062.

73 Baron Justus von Liebig published a monograph: Justus von Liebig and Lyon Playfair, *Organic Chemistry in Its Applications to Agriculture and Physiology* (London: Taylor and Walton, 1840).

73 modern synthetic fertilizer revolution: Greta Marchesi, "Justus von Liebig Makes the World: Soil Properties and Social Change in the Nineteenth Century," *Environmental Humanities* 12, no. 1 (2020): 205–26.

Chapter 4: Alive to Feeling

77 Touching an anesthetized person's body: André M. Bastos et al., "Neural effects of Propofol-Induced Unconsciousness and Its Reversal Using Thalamic Stimulation," *Elife* 10 (2021): e60824.

77 The drugs interfere with our action potentials: A. Taylor and G. McLeod, "Basic Pharmacology of Local Anaesthetics," *BJA Education* 20, no. 2 (2020): 34.

77 Venus flytraps under general anesthesia: Ken Yokawa et al., "Anaesthetics Stop Diverse Plant Organ Movements, Affect Endocytic Vesicle Recycling and ROS Ho-

meostasis, and Block Action Potentials in Venus Flytraps," *Annals of Botany* 122, no. 5 (2018): 747–56.

77   when mimosa is etherized: Thiago Paes de Barros De Luccia, "*Mimosa pudica, Dionaea muscipula* and anesthetics," *Plant Signaling and Behavior* 7, no. 9 (2012): 1163–67.

77   Brain waves decrease: S. Hagihira, "Changes in the Electroencephalogram during Anaesthesia and Their Physiological Basis," *British Journal of Anaesthesia* 115, suppl. 1 (2015): i27–i31.

78   a theory developed by neuroscientist Giulio Tononi: Giulio Tononi, "An Information Integration Theory of Consciousness," *BMC Neuroscience* 5 (2004): 1–22.

78   richness of this wave pattern: Carl Zimmer, "Sizing Up Consciousness by Its Bits," *New York Times*, September 20, 2010.

78   Slime mold . . . wavelike pulses: Gabriela Quirós, "This Pulsating Slime Mold Comes in Peace," *KQED*, April 19, 2016.

79   Mycelium . . . may coordinate: Elizabeth Gamillo, "Mushrooms May Communicate with Each Other Using Electrical Impulses," *Smithsonian Magazine*, April 2022.

79   without the need for a brain: Mirna Kramar and Karen Alim, "Encoding Memory in Tube Diameter Hierarchy of Living Flow Network," *Proceedings of the National Academy of Sciences* 118, no. 10 (2021): e2007815118.

79   He began his investigation by fastidiously stroking: Mordecai J. Jaffe, "Thigmomorphogenesis: The Response of Plant Growth and Development to Mechanical Stimulation: With Special Reference to *Bryonia dioica*," *Planta* 114 (1973): 143–57.

80   would grow girthier, and harden: Mordecai J. Jaffe, Frank W. Telewski, and Paul W. Cooke, "Thigmomorphogenesis: On the Mechanical Properties of Mechanically Perturbed Bean Plants," *Physiologia Plantarum* 62, no. 1 (1984): 73–78.

80   the same was true of Fraser firs: Frank W. Telewski and Mordecai J. Jaffe, "Thigmomorphogenesis: Field and Laboratory Studies of *Abies fraseri* in Response to Wind or Mechanical Perturbation," *Physiologia Plantarum* 66, no. 2 (1986): 211–18.

80   and loblolly pines: Frank W. Telewski and Mordecai J. Jaffe, "Thigmomorphogenesis: Anatomical, Morphological and Mechanical Analysis of Genetically Different Sibs of *Pinus taeda* in Response to Mechanical Perturbation," *Physiologia Plantarum* 66, no. 2 (1986): 219–26.

80   such a dramatic response in their hormones: Yue Xu et al., "Mitochondrial Function Modulates Touch Signalling in *Arabidopsis thaliana*," *Plant Journal* 97, no. 4 (2019): 623–45.

81   human touch . . . help plants ward off a future fungal infection: Lehcen Benikhlef et al., "Perception of Soft Mechanical Stress in *Arabidopsis* Leaves Activates Disease Resistance," *BMC Plant Biology* 13, no. 1 (2013): 1–12.

81   He hooked a cabbage to a voltmeter in front of the playwright George Bernard Shaw: Sir Patrick Geddes, *The Life and Work of Sir Jagadis C. Bose* (London: Longmans, Green, 1920), 146.

82   English scientist John Burdon-Sanderson: John Scott Burdon-Sanderson and F. J. M. Page, "I. On the Mechanical Effects and on the Electrical Disturbance Consequent on Excitation of the Leaf of *Dionæa muscipula*," *Proceedings of the Royal Society of London* 25, nos. 171–78 (1877): 411–34.

82   microelectrode system he designed himself: J. C. Bose, *The Nervous Mechanisms of Plants* (London: Longmans, Green, 1926), 184.

82   before scientists took the first microelectrode readings. . . in animals: Prakash Narain Tandon, "Jagdish Chandra Bose and Plant Neurobiology," *The Indian Journal of Medical Research* 149, no. 5 (2019): 593–599.

82   he wrote about "plant-nerves": Jagadis Chunder Bose and Guru Prasanna Das, "Physiological and Anatomical Investigations on *Mimosa pudica*," *Proceedings of the Royal Society of London B* 98, no. 690 (1925): 290–312.

82   "identical with that of the animal": J. C. Bose, *The Nervous Mechanism of Plants* (Calcutta: Longmans, Green, 1926), ix.

83   point-blank American racism: Peter V. Minorsky, "American Racism and the Lost Legacy of Sir Jagadis Chandra Bose, the Father of Plant Neurobiology," *Plant Signaling and Behavior* 16, no. 1 (2021): 1818030.

85   still accumulate the defensive proteins when another part of the plant was wounded: D. C. Wildon et al., "Electrical Signalling and Systemic Proteinase Inhibitor Induction in the Wounded Plant," *Nature* 360, no. 6399 (1992): 62–65.

86   mechanosensitive ion channels in plants: Jiu Ping Ding and Barbara G. Pickard, "Mechanosensory Calcium Selective Cation Channels in Epidermal Cells," *Plant Journal* 3, no. 1 (1993): 83–110.

86   studies on "stress in plants" with taxpayer money: Bill Clinton, "Remarks by the President in State of the Union Address," Washington, DC, 1995.

88   the flytrap can actually count: Jennifer Böhm et al., "The Venus Flytrap *Dionaea muscipula* Counts Prey-Induced Action Potentials to Induce Sodium Uptake," *Current Biology* 26, no. 3 (2016): 286–95.

93   A wave of green was moving across the plant: Masatsugu Toyota et al., "Glutamate triggers Long-Distance, Calcium-Based Plant Defense Signaling," *Science* 361, no. 6407 (2018): 1112–15.

95   glutamate-like receptors in plants would travel through the plant body: Seyed A. R. Mousavi et al., "Glutamate Receptor-Like Genes Mediate Leaf-to-Leaf Wound Signalling," *Nature* 500, no. 7463 (2013): 422–26.

96   triggering the adjacent cells to "freak out": Elizabeth Haswell and Ivan Baxter, "Simon Says: Captivate the Public with Snazzy Videos of Plant Defense, Send Plants to Space, and Embrace Curiosity-Driven Science," in *Taproot*, podcast, season 3, episode 5, March 19, 2019.

97   The lack of plant nerves didn't stop two scientific reviewers: Gloria K. Muday and Heather Brown-Harding, "Nervous System-Like Signaling in Plant Defense," *Science* 361, no. 6407 (2018): 1068–69.

97   a paper titled "Broadening the Definition": Sergio Miguel-Tomé and Rodolfo R. Llinás, "Broadening the Definition of a Nervous System to Better Understand the Evolution of Plants and Animals," *Plant Signaling and Behavior* 16, no. 10 (2021): 1927562.

99   still in its adolescence: Amber Dance, "The Quest to Decipher How the Body's Cells Sense Touch," *Nature* 577, no. 7789 (2020): 158–61.

Chapter 5: An Ear to the Ground

102   *Marcgravia evenia*, this ruby-colored sonar reflector: Ralph Simon et al., "Floral Acoustics: Conspicuous Echoes of a Dish-Shaped Leaf Attract Bat Pollinators," *Science* 333, no. 6042 (2011): 631–33.

102   *Mucuna holtonii* produces many small flowers: Dagmar von Helversen and Otto von Helversen, "Acoustic Guide in Bat-Pollinated Flower," *Nature* 398, no. 6730 (1999): 759–60.

107   the sound of its genuine predator chewing: Heidi M. Appel and Reginald B. Cocroft, "Plants Respond to Leaf Vibrations Caused by Insect Herbivore Chewing," *Oecologia* 175, no. 4 (2014): 1257–66.

108   playing arabidopsis a series of tones: Bosung Choi et al., "Positive Regulatory Role of Sound Vibration Treatment in *Arabidopsis thaliana* against *Botrytis cinerea* Infection," *Scientific Reports* 7, no. 1 (2017): 1–14.

108   playing some tones to rice for an hour: Mi-Jeong Jeong et al., "Sound Frequencies Induce Drought Tolerance in Rice Plant," *Pakistan Journal of Botany* 46 (2014): 2015–20.

108   increased the plants' content of vitamin C: Joo Yeol Kim et al., "Sound Waves Increases the ascorbic Acid Content of Alfalfa Sprouts by Affecting the Expression of Ascorbic Acid Biosynthesis-Related Genes," *Plant Biotechnology Reports* 11 (2017): 355–64.

108   increase their content of flavonoids: Joo Yeol Kim et al., "Sound Waves Affect the Total Flavonoid Contents in *Medicago sativa, Brassica oleracea* and *Raphanus sativus* Sprouts," *Journal of the Science of Food and Agriculture* 100, no. 1 (2020): 431–40.

109   the tiny hairs on arabidopsis leaves function as acoustic antennae: Shaobao Liu et al., "Arabidopsis Leaf Trichomes as Acoustic Antennae," *Biophysical Journal* 113, no. 9 (2017): 2068–76.

109   trichomes allow plants to sense the footsteps of moths and caterpillars: Michelle Peiffer et al., "Plants on Early Alert: Glandular Trichomes as Sensors for Insect Herbivores," *New Phytologist* 184, no. 3 (2009): 644–56.

110   Evening primrose . . . would increase the sweetness of its nectar: Marine Veits et al., "Flowers Respond to Pollinator Sound within Minutes by Increasing Nectar Sugar Concentration," *Ecology Letters* 22, no. 9 (2019): 1483–92.

111   nearly every pea plant grew its roots toward the sound of the running water: Monica Gagliano et al., "Tuned In: Plant Roots Use Sound to Locate Water," *Oecologia* 184, no. 1 (2017): 151–60.

111   Germany . . . spends an estimated 37 million euros per year repairing root-burst pipes: C. Bennerscheidt et al., "Unterirdische Infrastruktur–Bauteile, Bauverfahren and Schäden durch Wurzeln," in *Deutsche Baumpflegetage*, ed. D. Dujesiefken (Augsburg, Germany: Haymarket, 2009), 23–32 (in German). Cost adjusted to 2023 levels.

111   cause of more than half of all sewage pipe blockages: Thomas B. Randrup, E. Gregory McPherson, and Laurence R. Costello, "Tree Root Intrusion in Sewer Systems: Review of Extent and Costs," *Journal of Infrastructure Systems* 7, no. 1 (2001): 26–31.

112   Gagliano is urging her fellow researchers: Monica Gagliano, "Green Symphonies: A Call for Studies on Acoustic Communication in Plants," *Behavioral Ecology* 24, no. 4 (2013): 789–96.

112   "cavitation clicks" seem to increase . . . drought stress: Melvin T. Tyree and John S. Sperry, "Vulnerability of Xylem to Cavitation and Embolism," *Annual Review of Plant Biology* 40, no. 1 (1989): 19–36.

112   first solid evidence that the cavitation click theory could be true: Itzhak Khait et al., "Sounds Emitted by Plants under Stress Are Airborne and Informative," *Cell* 186, no. 7 (2023): 1328–36.

113   Gagliano compared the question to bat sonar: Monica Gagliano, "Green Symphonies: A Call for Studies on Acoustic Communication in Plants," *Behavioral Ecology* 24, no. 4 (2013): 789–96.

114   using echolocation to sense the position of the pole: Michael Pollan, "The Intelligent Plant," *New Yorker*, December 15, 2013. See also: Monica Gagliano, Michael Renton, Nili Duvdevani, Matthew Timmins, and Stefano Mancuso, "Acoustic and magnetic communication in plants: is it possible?." *Plant Signaling & Behavior* 7, no. 10 (2012): 1346-48.

114   prairie dogs appear to use adjectives: Leo Banks, "Scientist Has Gone to the Prairie Dogs, Finds They Talk," *Los Angeles Times*, June 5, 1997.

114   Japanese great tits have syntax: Toshitaka N. Suzuki, David Wheatcroft, and Michael Griesser, "Experimental Evidence for Compositional Syntax in Bird Calls," *Nature Communications* 7, no. 1 (2016): 10986.

115   radical pea-learning study: Monica Gagliano et al., "Learning by Association in Plants," *Scientific Reports* 6, no. 1 (2016): 38427.

115   the graduate student couldn't make it work: Kasey Markel, "Lack of Evidence for Associative Learning in Pea Plants," *Elife* 9 (2020): e57614.

116   a 2022 paper coauthored with University of California-Davis anthropologist Kristi Onzik: Kristi Onzik and Monica Gagliano, "Feeling Around for the Apparatus: A Radicley Empirical Plant Science," *Catalyst: Feminism, Theory, Technoscience 8*, no. 1 (2022), https://doi.org/10.28968/cftt.v8i1.34774.

Chapter 6: The (Plant) Body Keeps the Score

120　same ingredients to make their stinging hairs as humans and animals use to make their teeth: Hans-Jürgen Ensikat, Thorsten Geisler, and Maximilian Weigend, "A First Report of Hydroxylated Apatite as Structural Biomineral in Loasaceae–Plants' Teeth against Herbivores," *Scientific Reports* 6, no. 1 (2016): 26073.

120　pierce the skin of whatever animal ate them: Adeel Mustafa, Hans-Jürgen Ensikat, and Maximilian Weigend, "Stinging Hair Morphology and Wall Biomineralization across Five Plant Families: Conserved Morphology versus Divergent Cell Wall Composition," *American Journal of Botany* 105, no. 7 (2018): 1109–22.

121　"a flower that behaves like an animal" :"A Flower That Behaves Like an Animal," Freie Universität Berlin press release, August 12, 2012, https://www.fu-berlin.de/en/presse/informationen/fup/2012/fup_12_227/index.html.

121　it offers up larger globs of the sticky pollen at a time: Tilo Henning and Maximilian Weigend, "Total Control–Pollen Presentation and Floral Longevity in Loasaceae (Blazing Star Family) Are Modulated by Light, Temperature and Pollinator Visitation Rates," *PLoS ONE* 7, no. 8 (August 2012): e41121.

123　It was learning from experience: Moritz Mittelbach et al., "Flowers Anticipate Revisits of Pollinators by Learning from Previously Experienced Visitation Intervals," *Plant Signaling and Behavior* 14, no. 6 (2019): 1595320.

124　philosophers argue . . . that all memory shares a common basis with consciousness: Joachim Keppler, "The Common Basis of Memory and Consciousness: Understanding the Brain as a Write-Read Head Interacting with an Omnipresent Background Field," *Frontiers in Psychology* 10 (2020): 2968.

125　The body, it has been said, keeps the score: Bessel Van der Kolk, *The Body Keeps the Score: Brain, Mind, and Body in the Healing of Trauma* (New York: Penguin, 2014).

127　Cornish mallow . . . will turn its leaves hours before sunrise: Laura Ruggles, "The Minds of Plants," *Aeon*, December 12, 2017.

127　"extraordinarily complex–yet extremely elegant": Michael P. M. Dicker et al., "Biomimetic Photo-Actuation: Sensing, Control and Actuation in Sun-Tracking Plants," *Bioinspiration and Biomimetics* 9, no. 3 (2014): 036015.

128　to correct their own errors in judgment: Yuya Fukano, "Vine Tendrils Use Contact Chemoreception to Avoid Conspecific Leaves," *Proceedings of the Royal Society B* 284, no. 1850 (2017): 20162650.

129　"Plants in Motion": Roger P. Hangarter, Plants-In-Motion web page, https://plantsinmotion.bio.indiana.edu/.

129　sampling the air for the emanations of a suitable plant: Justin B. Runyon, Mark C. Mescher, and Consuelo M. De Moraes, "Volatile Chemical Cues Guide Host Location and Host Selection by Parasitic Plants," *Science* 313, no. 5795 (2006): 1964–67.

130　hawthorns that had been grown with extra nutrient supplements: Colleen K. Kelly, "Resource Choice in *Cuscuta europaea*," *Proceedings of the National Academy of Sciences* 89, no. 24 (1992): 12194–97.

130　The total number of coils: Anthony Trewavas, "The Foundations of Plant Intelligence," *Interface Focus* 7, no. 3 (2017): 20160098, section 10.3.

130　severe loss in twenty-five crop species across fifty-five countries: Bettina Kaiser et al., "Parasitic Plants of the Genus *Cuscuta* and Their Interaction with Susceptible and Resistant Host Plants," *Frontiers in Plant Science* 6 (2015): 45.

131　"compacted into a brain": Anthony Trewavas, "Intelligence, Cognition, and Language of Green Plants," *Frontiers in Psychology* 7 (2016): 588.

131　"Consciousness is thus not localized": Trewavas.

132　like stem cells, meristems are perpetually embryonic: Robin W. Kimmerer, "White Pine," in *The Mind of Plants: Narratives of Vegetal Intelligence* (Santa Fe, NM: Synergetic Press, 2021).

Chapter 7: Conversations with Animals

138　"rich wallow in multispecies muddles": Donna Haraway, "Tentacular Thinking:

Anthropocene, Capitalocene, Chthulucene," in *Staying with the Trouble: Making Kin in the Chthulucene* (Durham, NC: Duke University Press, 2016), 30–57.

139 De Moraes discovered this behavior in corn, tobacco, and cotton: Consuelo M. De Moraes et al., "Herbivore-Infested Plants Selectively Attract Parasitoids," *Nature* 393, no. 6685 (1998): 570–73.

140 bees biting plants made their flowers bloom as much as thirty days earlier: Foteini G. Pashalidou et al., "Bumble Bees Damage Plant Leaves and Accelerate Flower Production When Pollen Is Scarce," *Science* 368, no. 6493 (2020): 881–84.

141 The monkey flower is an excellent liar: Ariela I. Haber et al., "A Sensory Bias Overrides Learned Preferences of Bumblebees for Honest Signals in *Mimulus guttatus*," *Proceedings of the Royal Society B* 288, no. 1948 (2021): 20210161.

142 It's better, for the fly, to move on in search of a less-defended specimen: Eric C. Yip et al., "Sensory Co-Evolution: The Sex Attractant of a Gall-Making Fly Primes Plant Defences, but Female Flies Recognize Resulting Changes in Host-Plant Quality," *Journal of Ecology* 109, no. 1 (2021): 99–108.

143 The ants march the larvae deep into their ant nest: Tobias Lortzing et al., "Extrafloral Nectar Secretion from Wounds of *Solanum dulcamara*," *Nature Plants* 2, no. 5 (2016): 1–6.

143 symbiotic ants of the tropical tree genus *Macaranga*, who quickly die out when separated from it: Brigitte Fiala and Ulrich Maschwitz, "Studies on the South East Asian Ant-Plant Association *Crematogaster borneensis/Macaranga*: Adaptations of the Ant Partner," *Insectes sociaux* 37, no. 3 (1990): 212–31.

144 bacteria free-riding: E. Toby Kiers et al., "Host Sanctions and the Legume-Rhizobium Mutualism," *Nature* 425, no. 6953 (2003): 78–81.

146 "We could not have been more wrong": Rod Peakall, "Annals of Botany Lecture," filmed talk, July 28, 2020.

146 Almost all the semiochemicals he and his team analyzed were entirely new to plant science: Rod Peakall, "Q&A: Rod Peakall," *Current Biology Magazine* 32, no. 16 (2022): R861–R863. https://www.cell.com/current-biology/pdf /S09609822(22)01129-0.pdf>

146 two or more of these compounds: Haiyang Xu et al., "Complex Sexual Deception in an Orchid Is Achieved by Co-opting Two Independent Biosynthetic Pathways for Pollinator Attraction," *Current Biology* 27, no. 13 (2017): 1867–77.

147 offer a different way of seeing the relationship between orchid and wasp: Carla Hustak and Natasha Myers, "Involutionary Momentum: Affective Ecologies and the Sciences of Plant/Insect Encounters," *differences* 23, no. 3 (2012): 74–118.

147 The most perfect adaptation in nature: Hustak and Myers (2012): p. 74, Darwin quoted therein: "In no other plant, or indeed in hardly any animal, can adaptations of one part to another, and of the whole to other organisms widely remote in the scale of nature, be named more perfect than those presented by this Orchis."

148 "inextricable web of affinities": Charles Darwin, *On the Origin of Species*, 1866.

148 not absolutely indistinguishable from the real thing: Nicolas J. Vereecken and Florian P. Schiestl, "The Evolution of Imperfect Floral Mimicry," *Proceedings of the National Academy of Sciences* 105, no. 21 (2008): 7484–88.

148 why asters and goldenrod tended to bloom together each September: Robin W. Kimmerer, "Asters and Goldenrod," in *Braiding Sweetgrass: Indigenous Wisdom, Scientific Knowledge and the Teachings of Plants* (Minneapolis: Milkweed, 2013).

149 Flowers themselves evolved in order to be beautiful to animals: Ferris Jabr, "How Beauty Is Making Scientists Rethink Evolution," *New York Times Magazine*, January 9, 2019.

151 spontaneously switch the sex of a section of its body: Toshiyuki Nagata et al., "Sex Conversion in Ginkgo biloba (Ginkgoaceae)," *Journal of Japanese Botany* 91 (2016): 120–27.

153 they described plants' biochemical synthesis as a "language": Jarmo K. Holopainen

and James D. Blande, "Molecular Plant Volatile Communication," in *Sensing in Nature*, ed. Carlos López Larrea (New York: Springer, 2012), 17–31.

153   The silver birch absorbed the scent from its plant neighbor: Sari J. Himanen et al., "Birch (*Betula* spp.) leaves Adsorb and Re-release Volatiles Specific to Neighbouring Plants–A Mechanism for Associational Herbivore Resistance?," *New Phytologist* 186, no. 3 (2010): 722–32.

154   The pollution steadily filling the air: Jarmo K. Holopainen, Anne-Marja Nerg, and James D. Blande, "Multitrophic Signalling in Polluted Atmospheres," in *Biology, Controls and Models of Tree Volatile Organic Compound Emissions*, ed. Ülo Niinemets and Russell K. Monson (Dordrecht, Germany: Springer, 2013), 285–314.

155   when black mustard flowers are exposed to ozone: Gerard Farré-Armengol et al., "Ozone Degrades Floral Scent and Reduces Pollinator Attraction to Flowers," *New Phytologist* 209, no. 1 (2016): 152–60.

156   entirely unable to summon beneficial predators: Amanuel Tamiru et al., "Maize Landraces Recruit Egg and Larval Parasitoids in Response to Egg Deposition by a Herbivore," *Ecology Letters* 14, no. 11 (2011): 1075–83.

156   We are clearly losing the war on pests: Kat McGowan, "Listen to the Plants," *Slate*, April 18, 2014.

156   two million tons of conventional pesticides: Anket Sharma et al., "Worldwide Pesticide Usage and Its Impacts on Ecosystem," *SN Applied Sciences* 1, no. 11 (2019): 1–16.

156   The United States alone says it uses one billion pounds a year: "Pesticides," Pesticides webpage, U.S. Geological Survey, 2017. https://www.usgs.gov/centers/ohio-kentucky-indiana-water-science-center/science/pesticides?qt-science_center_objects=0#overview.

156   11,000 farmworkers are fatally poisoned by pesticides each year: Wolfgang Boedeker et al., "the global distribution of acute unintentional pesticide poisoning: estimations based on a systematic review," *BMC Public Health* 20, no. 1 (2020): 1–19.

157   breed rice plants to include a terpene from lima beans: Fengqi Li et al., "Expression of Lima Bean Terpene Synthases in Rice Enhances Recruitment of a Beneficial Enemy of a Major Rice Pest," *Plant, Cell and Environment* 41, no. 1 (2018): 111–20.

157   Some plant scientists argue for exploiting more of plants' natural defense mechanisms: Mirian F. F. Michereff et al., "Variability in Herbivore-Induced Defence Signalling across Different Maize Genotypes Impacts Significantly on Natural Enemy Foraging Behaviour," *Journal of Pest Science* 92 (2019): 723–36.

157   Farmers know that strawberries will produce a third more fruit: Janine Griffiths-Lee, Elizabeth Nicholls, and Dave Goulson, "Companion Planting to Attract Pollinators Increases the Yield and Quality of Strawberry Fruit in Gardens and Allotments," *Ecological Entomology* 45, no. 5 (2020): 1025–34.

157   The borage attracts the strawberry's pollinator: Nathan Hecht, "Berries, Bees, and Borage," Minnesota Fruit Research, University of Minnesota, December 3, 2018, https://fruit.umn.edu/content/berries-bees-borage.

Chapter 8: The Scientist and the Chameleon Vine

160   Darwin, was himself absorbed: Charles Darwin, *The Movements and Habits of Climbing Plants*, 2nd ed. (London: John Murray, 1875).

163   grow ten times longer in a laboratory: Ken Yokawa, Tomoko Kagenishi, and František Baluška, "Root Photomorphogenesis in Laboratory-Maintained Arabidopsis Seedlings," *Trends in Plant Science* 18, no. 3 (2013): 117–19.

163   actually just running away: Yokawa, Kagenishi, and Baluška.

163   shown light phobia in corn: Christian Burbach et al., "Photophobic Behavior of Maize Roots," *Plant Signaling and Behavior* 7, no. 7 (2012): 874–78.

164   Vavilov discovered a strange phenomenon: J. Scott McElroy, "Vavilovian Mimicry:
      Nikolai Vavilov and His Little-Known Impact on Weed Science," *Weed Science* 62,
      no. 2 (2014): 207–16.
165   began to change its architecture to match the rice: C. Y. Ye et al., Genomic Evidence
      of Human Selection on Vavilovian Mimicry, *Nature Ecology and Evolution* 3, no.
      10 (2019): 1474–82.
165   Vavilovian mimicry at the biochemical level: McElroy, "Vavilovian Mimicry."
167   "Vision in Plants via Plant-Specific Ocelli?": František Baluška and Stefano Man-
      cuso, "Vision in Plants via Plant-Specific Ocelli?," *Trends in Plant Science* 21, no. 9
      (2016): 727–30.
167   *The Light-Sensing Organs of Leaves*: Gottlieb Haberlandt, *Die Lichtsinnesorgane
      der Laubblätter* (Leipzig, Germany: W. Engelmann, 1905).
167   Francis Darwin . . . referenced it at length: Francis Darwin, "Lectures on the Physi-
      ology of Movement in Plants," *New Phytologist* 5, no. 9 (November 1906): 74.
168   Their cells act as "spherical microlenses": Nils Schuergers et al., "Cyanobacteria
      Use Micro-Optics to Sense Light Direction," *Elife* 5 (2016): e12620.
169   parasitic plants can read this changing light ratio: Jason D. Smith et al., "A Plant
      Parasite Uses Light Cues to Detect Differences in Host Plant Proximity and Archi-
      tecture," *Plant, Cell and Environment* 44, no. 4 (2021): 1142–50.
169   fourteen types of light receptors in plants: Inyup Paik and Enamul Huq, "Plant
      Photoreceptors: Multi-Functional Sensory Proteins and Their Signaling Net-
      works," *Seminars in Cell and Developmental Biology* 92 (2019): 114–21.
169   In one 2014 paper, botanists in Argentina: María A. Crepy and Jorge J. Casal,
      "Photoreceptor Mediated Kin Recognition in Plants," *New Phytologist* 205, no. 1
      (2015): 329–38.
171   Published a suite of remarkable boquila findings: Ernesto Gianoli and Fernando
      Carrasco-Urra, "Leaf Mimicry in a Climbing Plant Protects against Herbivory,"
      Current Biology 24, no. 9 (2014): 984–87.
172   a sprig of she-oak mistletoe beside an Australian river oho oak: Bryan Barlow,
      "Cryptic Mimicry of Their Hosts–Mistletoes," Australian National Herbarium,
      September 11, 2012, https://www.anbg.gov.au/mistletoe/mimicry.html.
182   2,600 distinct types of micro RNA: Olga Plotnikova, Ancha Baranova, and Mikhail
      Skoblov, "Comprehensive Analysis of Human MicroRNA–mRNA Interactome,"
      Frontiers in Genetics 10 (2019): 933.
182   one-third of the genes in our genome: Scott M. Hammond, "An Overview of Mi-
      croRNAs," *Advanced Drug Delivery Reviews* 87 (2015): 3–14.
182   The small RNA from one plant . . . interfere with the gene expression in other, nearby
      plants: Federico Betti et al., "Exogenous miRNAs Induce Post-Transcriptional
      Gene Silencing in Plants," *Nature Plants* 7, no. 10 (2021): 1379–88.
183   termites were recently discovered to have microbes in their guts that make it pos-
      sible: Kazuki Izawa et al., "Discovery of Ectosymbiotic Endomicrobium lineages
      Associated with Protists in the Gut of Stolotermitid Termites," *Environmental Mi-
      crobiology Reports* 9, no. 4 (2017): 411–18.
184   Non-mimicking leaves . . . had a totally different bacterial community: Ernesto Gi-
      anoli et al.,"Endophytic Bacterial Communities Are Associated with Leaf Mimicry
      in the Vine *Boquila trifoliolata*," *Scientific Reports* 11, no. 1 (2021): 22673.
185   Rupert Sheldrake's concept of a "morphogenetic field": Rupert Sheldrake, "Mor-
      phic Resonance and Morphic Fields–An Introduction," https://www.sheldrake
      .org/research/morphic-resonance/introduction.
185   "You know the dirty kid from *Peanuts*? Pig-Pen?": Zoë Schlanger, "Your Microbi-
      ome Extends in a Microbial Cloud Around You, Like an Aura," *Newsweek*, Septem-
      ber 22, 2015.
186   Our own cells are likely outnumbered by our microbial tenants: Ron Sender,
      Shai Fuchs, and Ron Milo, "Are We Really Vastly Outnumbered? Revisiting the
      Ratio of Bacterial to Host Cells in Humans," *Cell* 164, no. 3 (2016): 337–40. See

also James Gallagher, "More Than Half Your Body Is Not Human," *BBC News* 251 (2018).

188   Lynn Margulis first popularized the concept of a "holobiont": Jean-Christophe Si-mon et al., "Host-Microbiota Interactions: From Holobiont Theory to Analysis," *Microbiome* 7, no. 1 (2019): 1–5.

188   rejected by fifteen journals: Bruce Weber, "Lynn Margulis, Evolution Theorist, Dies at 73," *New York Times*, November 24, 2011.

188   It was proven true a decade later: Michael W. Gray, Gertraud Burger, and B. Franz Lang, "Mitochondrial Evolution," *Science* 283, no. 5407 (1999): 1476–81.

189   result of microbial signals: Thomas C. G. Bosch and Margaret McFall-Ngai, "An-imal Development in the Microbial World: Re-Thinking the Conceptual Frame-work," *Current Topics in Developmental Biology* 141 (2021): 399–427.

189   "A memory-based immune system may have evolved": Margaret McFall-Ngai, "Care for the Community," *Nature* 445, no. 7124 (2007): 153.

189   "People and other eukaryotes are like solids frozen": Lynn Margulis and Dorion Sagan, *Microcosmos: Four Billion Years of Microbial Evolution* (Berkeley: Univer-sity of California Press, 1997).

192   "The completely self-contained 'individual' is a myth": Lynn Margulis and Dorion Sagan, *Acquiring Genomes: A Theory of the Origin of Species* (New York: Basic Books, 2008).

Chapter 9: The Social Life of Plants

193   An entomologist named this lifestyle "eusocial" behavior: Suzanne Batra, "Nests and Social Behavior of Halictine Bees of India (Hymenoptera: Halictidae)," *Indian Journal of Entomology* 28 (1966): 375.

193   *Eusocial* literally means "truly social": Suzanne Batra, "Beyond the Honeybee," *American Scientist* 110, no. 2 (2022): 72–74.

194   eusocial behavior must have evolved separately many times: Michael R. Warner et al., "Convergent Eusocial Evolution Is Based on a Shared Reproductive Groundplan plus Lineage-Specific Plastic Genes," *Nature Communications* 10, no. 1 (2019): 2651.

194   Could plants be eusocial too: K. C. Burns, Ian Hutton, and Lara Shepherd, "Primi-tive Eusociality in a Land Plant?," *Ecology* 102, no. 9 (2021): e03373.

195   electrical activity in the human brain can synchronize between people: Sivan Kinreich et al., "Brain-to-Brain Synchrony during Naturalistic Social Interactions," *Scientific Reports* 7, 17060 (December 2017).

195   That type of synchronization has been found in bats: Julia Sliwa, "Toward Collec-tive Animal Neuroscience," *Science* 374, no. 6566 (October 2021).

195   people perform better when their brainwaves are synchronized: Caroline Szyman-ski et al., "Teams on the Same Wavelength Perform Better: Inter-Brain Phase Syn-chronization Constitutes a Neural Substrate for Social Facilitation," *Neuroimage* 152 (2017): 425–436.

195   copilots tend to synchronize during takeoff: Laura Astolfi et al., "Cortical Activity and Functional Hyperconnectivity by Simultaneous EEG Recordings from Inter-acting Couples of Professional Pilots," in *Proceedings of the Annual International Conference of the IEEE Engineering in Medicine and Biology Society* (New York: IEEE, 2012), 4752–55.

195   higher feelings of cooperativeness: Yi Hu et al., "Brain-to-Brain Synchronization across Two Persons Predicts Mutual Prosociality," *Social Cognitive and Affective Neuroscience* 12, no. 12 (2017): 1835–44.

195   Couples . . . report more satisfaction: Lei Li et al., "Neural Synchronization Predicts Marital Satisfaction," *Proceedings of the National Academy of Sciences* 119, no. 34 (2022): e2202515119.

195   brains of co-parents appear to sync up: Atiqah Azhari et al., "Physical Presence of Spouse Enhances Brain-to-Brain Synchrony in Co-parenting Couples," *Scientific Reports* 10, no. 1 (2020): 1–11.

196   plants know exactly who their siblings are: Susan A. Dudley and Amanda L. File, "Kin Recognition in an Annual Plant," *Biology Letters* 3, no. 4 (2007): 435–38.

199   When planted beside kin: Guillermo P. Murphy and Susan A. Dudley, "Kin Recognition: Competition and Cooperation in Impatiens (Balsaminaceae)," *American Journal of Botany* 96, no. 11 (2009): 1990–96.

199   Hamilton's Rule states that you will behave preferentially toward your family: Andy Gardner and Stuart A. West, "Inclusive Fitness: 50 Years On," *Philosophical Transactions of the Royal Society B* 369, no. 1642 (2014): 20130356.

199   Haldane was rumored to have declared: Gardner and West.

199   orca whales live in complex family pods: Lisa Stiffler, "Understanding Orca Culture," *Smithsonian Magazine*, August 2011.

199   female baboons spend their entire lives within yards of their mothers: "Baboon Social Life," Amboseli Baboon Research Project, Princeton University, https://www.princeton.edu/~baboon/social_life.html.

200   sponge-dwelling shrimp are known to collaborate with their family members: Emmett J. Duffy, Cheryl L. Morrison, and Kenneth S. Macdonald, "Colony Defense and Behavioral Differentiation in the Eusocial Shrimp *Synalpheus regalis*," *Behavioral Ecology and Sociobiology* 51 (2002): 488–95.

200   sunflower farmers could get up to 47 percent more oil yield: Mónica López Pereira et al., "Light-Mediated Self-Organization of Sunflower Stands Increases Oil Yield in the Field," *Proceedings of the National Academy of Sciences* 114, no. 30 (2017): 7975–80.

200   how sagebrush in California defend themselves: Richard Karban et al., "Kin recognition Affects Plant Communication and Defence," *Proceedings of the Royal Society B* 280, no. 1756 (2013): 20123062.

200   An arabidopsis plant will rearrange its leaves: María A. Crepy and Jorge J. Casal, "Photoreceptor Mediated Kin Recognition in Plants," *New Phytologist* 205, no. 1 (2015): 329–38.

201   carnivorous plants . . . have evolved to hunt in packs: Kazuki Tagawa and Mikio Watanabe, "Group Foraging in Carnivorous Plants: Carnivorous Plant *Drosera makinoi* (Droseraceae) Is More Effective at Trapping Larger Prey in Large Groups," *Plant Species Biology* 36, no. 1 (2021): 114–18.

201   The team grew over a dozen different rice lines: Xue Fang Yang et al., "Kin Recognition in Rice (*Oryza sativa*) lines," *New Phytologist* 220, no. 2 (2018): 567–78.

203   invest more in advertising to pollinators: Rubén Torices, José M. Gómez, and John R. Pannell, "Kin Discrimination Allows Plants to Modify Investment towards Pollinator Attraction," *Nature Communications* 9, no. 1 (2018).

204   Dudley suggested that crop breeders have been going about their business with a huge blind spot: Guillermo P. Murphy, Clarence J. Swanton, Rene C. Van Acker, and Susan A. Dudley. "Kin Recognition, Multilevel Selection and Altruism in Crop Sustainability," *Journal of Ecology* 105, no. 4 (2017): 930–934.

205   Akira Yamawo . . . tested this capacity on Asiatic plantains: Akira Yamawo and Hiromi Mukai, "Seeds Integrate Biological Information about Conspecific and Allospecific Neighbours," *Proceedings of the Royal Society B* 284, no. 1857 (2017): 20170800.

206   counted the roots of a single winter rye plant: Howard J. Dittmer, "A Quantitative Study of the Roots and Root Hairs of a Winter Rye Plant (*Secale cereale*)," *American Journal of Botany* (1937): 417–20.

206   The amino acids glutamate and glycine . . . pass between plants and fungi: Suzanne W. Simard, "Mycorrhizal Networks Facilitate Tree Communication, Learning, and Memory," in *Memory and Learning in Plants*, ed. F. Baluška, M. Gagliano, and G. Witzany (New York: Springer, 2018), 191–213.

206   *Entangled Life*: Merlin Sheldrake, *Entangled Life: How Fungi Make Our Worlds, Change Our Minds and Shape Our Futures* (New York: Random House, 2021).

207   But suddenly the dry-land grass did just fine in brine: Zoë Schlanger, "Our Silent Partners," *New York Review of Books,* October 7, 2021.

207   The sweetness of tomatoes: A. Copetta et al., "Fruit Production and Quality of Tomato Plants (*Solanum lycopersicum* L.) Are Affected by Green Compost and Arbuscular Mycorrhizal Fungi," *Plant Biosystems* 145, no. 1 (2011): 106–115.

207   the aromatic qualities of basil: A. Copetta, G. Lingua, and G. Berta, "Effects of Three AM Fungi on Growth, Distribution of Glandular Hairs, and Essential Oil Production in *Ocimum basilicum* L. var. *Genovese," Mycorrhiza* 16 (2006): 485–94.

207   The concentrations of medicinal compounds in echinacea: Ghada Araim et al., "Root Colonization by an Arbuscular Mycorrhizal (AM) Fungus Increases Growth and Secondary Metabolism of Purple Coneflower, *Echinacea purpurea* (L.) Moench," *Journal of Agricultural and Food Chemistry* 57, no. 6 (2009): 2255–58.

207   the aromatics in patchouli: J. Arpana et al., "Symbiotic Response of Patchouli [*Pogostemon cablin* (Blanco) Benth.] to Different Arbuscular Mycorrhizal Fungi," *Advances in Environmental Biology* 2, no. 1 (2008): 20–24.

207   the antioxidant power of artichoke heads: Nello Ceccarelli et al., "Mycorrhizal Colonization Impacts on Phenolic Content and Antioxidant Properties of Artichoke Leaves and Flower Heads Two Years after Field Transplant," *Plant and Soil* 335 (2010): 311–23.

207   roots are . . . made to stitch plants and fungi into relation: Sheldrake, *Entangled Life.*

208   plants can preferentially direct carbon: E. Toby Kiers et al., "Reciprocal Rewards Stabilize Cooperation in the Mycorrhizal Symbiosis," *Science* 333, no. 6044 (2011): 880–82.

208   The swarming ability of roots: Marzena Ciszak et al., "Swarming Behavior in Plant Roots," *PLoS One* 7, no. 1 (2012): e29759.

209   Each root tip acts as both a gatherer and a sensor: Suqin Fang et al., "Genotypic Recognition and Spatial Responses by Rice Roots," *Proceedings of the National Academy of Sciences* 110, no. 7 (2013): 2670–75.

209   that roots actively forage for their food: James F. Cahill Jr. and Gordon G. McNickle, "The Behavioral Ecology of Nutrient Foraging by Plants," *Annual Review of Ecology, Evolution, and Systematics* 42 (2011): 289–311.

209   *behavior* is also a term he prefers to use whenever possible: James F. Cahill, "Introduction to the Special Issue: Beyond Traits: Integrating Behaviour into Plant Ecology and Biology," *AoB Plants* 7 (2015). See also James F. Cahill Jr., "The Inevitability of Plant Behavior," *American Journal of Botany* 106, no. 7 (2019): 903–5.

209   bad foraging decisions: Akira Yamawo, Haruna Ohsaki, and James F. Cahill Jr., "Damage to Leaf Veins Suppresses Root Foraging Precision," *American Journal of Botany* 106, no. 8 (2019): 1126–30.

210   people make poorer decisions when stressed: Jordan Skrynka and Benjamin T. Vincent, "Hunger Increases Delay Discounting of Food and Non-Food Rewards," *Psychonomic Bulletin and Review* 26, no. 5 (2019): 1729–37.

210   sunflowers will take note of their social environment: Megan K. Ljubotina and James F. Cahill Jr., "Effects of Neighbour Location and Nutrient Distributions on Root Foraging Behaviour of the Common Sunflower," *Proceedings of the Royal Society B* 286, no. 1911 (2019): 20190955.

Chapter 10: Inheritance

215   The plant, *Spigelia genuflexa,* has planted its own seeds: Alex V. Popovkin et al., "*Spigelia genuflexa* (Loganiaceae), a New Geocarpic Species from the Atlantic Forest of Northeastern Bahia, Brazil," *PhytoKeys* 6 (2011): 47.

215   "common human activity": "Amateur Botanists Discover a Genuflecting Plant in Brazil," Rutgers University, September 18, 2011, https://www.rutgers.edu/news /amateur-botanists-discover-genuflecting-plant-brazil.

217   plantain lightens the color of the spike: Elizabeth P. Lacey and David Herr, "Phenotypic Plasticity, Parental Effects, and Parental Care in Plants? I. An Examination

of Spike Reflectance in *Plantago lanceolata* (Plantaginaceae)," *American Journal of Botany* 92, no. 6 (2005): 920–30.

217 If a parent plant finds itself in a drier environment: Sonia E. Sultan, *Organism and Environment: Ecological Development, Niche Construction, and Adaptation* (New York: Oxford University Press, 2015), 88, and references therein.

217 In the high alpine ridges of Colorado: Anna Wied and Candace Galen, "Plant Parental Care: Conspecific Nurse Effects in *Frasera speciosa* and *Cirsium scopulorum*," *Ecology* 79, no. 5 (1998): 1657–68.

217 if yellow monkey flowers are exposed to predators: Alison G. Scoville et al., "Differential Regulation of a MYB Transcription Factor Is Correlated with Transgenerational Epigenetic Inheritance of Trichome Density in *Mimulus guttatus*," *New Phytologist* 191, no. 1 (2011): 251–63.

217 Wild radishes that have lived through a scourge: Anurag A. Agrawal, Christian Laforsch, and Ralph Tollrian, "Transgenerational Induction of Defences in Animals and Plants," *Nature* 401, no. 6748 (1999): 60–63.

223 a paragraph on a U. S. government website about the human genome: "Genomics and Its Impact on Science and Society," U.S. Department of Energy Genome Research Programs, http://www.ornl.gov/sci/techresources/Human_Genome /publicat/primer2001/primer11.pdf

223 "The big A and the little a, the gene for tall and the gene for short": "Extending Evolution, an Interview with Prof. Sonia Sultan," episode 60 of *Naturally Speaking*, podcast, April 2018.

223 inheritance seems to explain only about 36 percent of the heritability of a person's height: Sonia E. Sultan, Armin P. Moczek, and Denis Walsh, "Bridging the Explanatory Gaps: What Can We Learn from a Biological Agency Perspective?," *BioEssays* 44, no. 1 (2022): 2100185.

224 more surface area to catch falling photons: Sonia E. Sultan, "Plant Developmental Responses to the Environment: Eco-Devo Insights," *Current Opinion in Plant Biology* 13, no. 1 (2010): 96–101.

224 goldfish... can completely remodel their gills: Jørund Sollid and Göran E. Nilsson, "Plasticity of Respiratory Structures—Adaptive Remodeling of Fish Gills Induced by Ambient Oxygen and Temperature," *Respiratory Physiology and Neurobiology* 154, no. 1–2 (2006): 241–51.

225 morphs into deep-rooted, long-rooted seedlings: Jacob J. Herman et al., "Adaptive Transgenerational Plasticity in an Annual Plant: Grandparental and Parental Drought Stress Enhance Performance of Seedlings in Dry Soil," *Integrative and Comparative Biology* 52, no. 1 (July 2012): 77–88.

225 a study from 2000, on smoking, genes, broccoli, and lung cancer: Margaret R. Spitz et al., "Dietary Intake of Isothiocyanates: Evidence of a Joint Effect with Glutathione S-transferase Polymorphisms in Lung Cancer Risk," *Cancer Epidemiology Biomarkers and Prevention* 9, no. 10 (2000): 1017–20.

225 it even appeared to help people without the genetic anomaly: Julie E. Bauman et al., "Randomized Crossover Trial Evaluating Detoxification of Tobacco Carcinogens by Broccoli Seed and Sprout Extract in Current Smokers," *Cancers* 14, no. 9 (2022): 2129.

226 "On closer examination, however, the environment extends": Sultan, *Organism and Environmnent*, 31.

226 Consider the emerald green sea slug: Mary E. Rumpho et al., "The Making of a Photosynthetic Animal," *Journal of Experimental Biology* 214, no. 2 (2011): 303–11.

227 The slug orients its body in the same way a leaf does: Sultan, *Organism and Environment*, 32.

228 plants exist in a state of total "immersion": Emanuele Coccia, *The Life of Plants: A Metaphysics of Mixture* (Hoboken, NJ: John Wiley, 2019).

229 A few years prior, I'd gone to Detroit: Zoë Schlanger, "Choking to Death in Detroit: Flint Isn't Michigan's Only Disaster," *Newsweek*, March 30, 2016.

230 dozens of diseases . . . fall under the missing heritability problem: Teri A. Manolio et al., "Finding the Missing Heritability of Complex Diseases," *Nature* 461, no. 7265 (2009): 747–53.

232 On Sultan's lab website, it says the team studies plant "monsters": "Research: Current Projects," Sultan Lab, Wesleyan University. https://sultanlab.research .wesleyan.edu/currentprojects/.

233 smartweed plants whose parents had to compete against neighbors for light: Robin Waterman and Sonia E. Sultan, "Transgenerational Effects of Parent Plant Competition on Offspring Development in Contrasting Conditions," *Ecology* 102, no. 12 (2021): e03531.

234 They will have a leg up on their peers: Brennan H. Baker et al., "Transgenerational Effects of Parental Light Environment on Progeny Competitive Performance and Lifetime Fitness," *Philosophical Transactions of the Royal Society B* 374, no. 1768 (2019): 20180182.

234 collectors seeking to add an attractive exotic species: Peter Del Tredici, "The Introduction of Japanese Knotweed, *Reynoutria japonica*, into North America," *Journal of the Torrey Botanical Society* 144, no. 4 (2017): 406–16.

237 growing within three meters: Philip Santo, "New Japanese Knotweed Standard Comes into Effect," *Property Journal*, RICS, March 21, 2022.

237 National Park Service anxiously watch knotweed grow: Sophia Cameron, "Invasive Plant Profile: Japanese Knotweed," Acadia National Park, National Park Service, https://www.nps.gov/articles/000/japanese-knotweed-acadia.htm

237 miles upon unbroken miles of the plant growing along the Bronx River: David Taft, "Japanese Knotweed Is Here to Stay," *New York Times*, September 6, 2018.

238 The idea that plants have agency: Sonia E. Sultan, Armin P. Moczek, and Denis Walsh, "Bridging the Explanatory Gaps: What Can We Learn from a Biological Agency Perspective?," *BioEssays* 44, no. 1 (2022): 2100185.

Chapter 11: Plant Futures

246 The philosopher Bruno Latour once wrote: Bruno Latour, "A Collective of Humans and Nonhumans: Following," *Readings in the Philosophy of Technology* (2009): 156.

249 Some, like ecologist Carl Safina, argue: C. A. Safina, "Why Anthropomorphism Helps Us Understand Animals' Behavior," Medium.com, September 9, 2016.

249 "It is by the help of the better known": Theophrastus, *Historia Plantarum* 1.2.5.

249 anthropologist Natasha Myers noted: Natasha Myers, "Conversations on Plant Sensing: Notes from the Field," *NatureCulture* 3 (2015): 35–66.

251 Jeffrey T. Nealon wonders: Jeffrey T. Nealon, *Plant Theory: Biopower and Vegetable Life* (Stanford, CA: Stanford University Press, 2015).

251 made it to the Supreme Court: Supreme Court of United States, *Sierra Club v. Morton*, 405 U.S. 727 (1972).

252 "Should Trees Have Standing?": Christopher D. Stone, "Should Trees Have Standing? Toward Legal Rights for Natural Objects," S. Cal. l. rev. 45 (1972): 450.

253 wild rice sued the state of Minnesota: *Manoomin v. Minnesota Department of Natural Resources*, case no. GC21-0428, White Earth Band of Ojibwe Tribal Ct. (2021).

253 "inherent rights to exist, flourish, regenerate, and evolve": "Rights of Manoomin," section 1, White Earth Reservation Business Committee, White Earth Band of Chippewa Indians, resolution no. 001-19-009, December 31, 2018.

254 "intersubjective encounter": Deborah Bird Rose, "Indigenous Ecologies and an Ethic of Connection," in *Global Ethics And Environment*, ed. Nicholas Low (London: Routledge, 1999), 175.

255 wrote about this in-betweenness: Báyò Akómoláfé, "When You Meet the Monster, Anoint Its Feet," *Emergence Magazine*, October 16, 2018.

Index